# Nuclear Physics
## Exploring the Heart of Matter

Committee on the Assessment of and Outlook for Nuclear Physics

Board on Physics and Astronomy

Division on Engineering and Physical Sciences

NATIONAL RESEARCH COUNCIL
*OF THE NATIONAL ACADEMIES*

THE NATIONAL ACADEMIES PRESS
Washington, D.C.
**www.nap.edu**

**THE NATIONAL ACADEMIES PRESS    500 Fifth Street, NW    Washington, DC 20001**

NOTICE: The project that is the subject of this report was approved by the Governing Board of the National Research Council, whose members are drawn from the councils of the National Academy of Sciences, the National Academy of Engineering, and the Institute of Medicine. The members of the committee responsible for the report were chosen for their special competences and with regard for appropriate balance.

This study was supported by Grant No. PHY-80933 between the National Academy of Sciences and the National Science Foundation and by Grant No. DE-SC0002593 between the National Academy of Sciences and the Department of Energy. Any opinions, findings, conclusions, or recommendations expressed in this publication are those of the author(s) and do not necessarily reflect the views of the organizations or agencies that provided support for the project.

*Cover:* "Circles in a Circle" by Wassily Kandinsky.

*Dedication* (p. xv): Photo courtesy of University of California, Berkeley, Department of Physics.

International Standard Book Number-13:    978-0-309-0-309-26040-4
International Standard Book Number-10:    0-309-26040-X
Library of Congress Control Number:    2013931504

Additional copies of this report are available from the National Academies Press, 500 Fifth Street, NW, Keck 360, Washington, DC 20001; (800) 624-6242 or (202) 334-3313; http://www.nap.edu; and the Board on Physics and Astronomy, National Research Council, 500 Fifth Street, NW, Washington, DC 20001; http://www.national-academies.org/bpa.

# THE NATIONAL ACADEMIES
*Advisers to the Nation on Science, Engineering, and Medicine*

The National Academy of Sciences is a private, nonprofit, self-perpetuating society of distinguished scholars engaged in scientific and engineering research, dedicated to the furtherance of science and technology and to their use for the general welfare. Upon the authority of the charter granted to it by the Congress in 1863, the Academy has a mandate that requires it to advise the federal government on scientific and technical matters. Dr. Ralph J. Cicerone is president of the National Academy of Sciences.

The National Academy of Engineering was established in 1964, under the charter of the National Academy of Sciences, as a parallel organization of outstanding engineers. It is autonomous in its administration and in the selection of its members, sharing with the National Academy of Sciences the responsibility for advising the federal government. The National Academy of Engineering also sponsors engineering programs aimed at meeting national needs, encourages education and research, and recognizes the superior achievements of engineers. Dr. Charles M. Vest is president of the National Academy of Engineering.

The Institute of Medicine was established in 1970 by the National Academy of Sciences to secure the services of eminent members of appropriate professions in the examination of policy matters pertaining to the health of the public. The Institute acts under the responsibility given to the National Academy of Sciences by its congressional charter to be an adviser to the federal government and, upon its own initiative, to identify issues of medical care, research, and education. Dr. Harvey V. Fineberg is president of the Institute of Medicine.

The National Research Council was organized by the National Academy of Sciences in 1916 to associate the broad community of science and technology with the Academy's purposes of furthering knowledge and advising the federal government. Functioning in accordance with general policies determined by the Academy, the Council has become the principal operating agency of both the National Academy of Sciences and the National Academy of Engineering in providing services to the government, the public, and the scientific and engineering communities. The Council is administered jointly by both Academies and the Institute of Medicine. Dr. Ralph J. Cicerone and Dr. Charles M. Vest are chair and vice chair, respectively, of the National Research Council.

**www.national-academies.org**

# Preface

The National Research Council convened the Committee on the Assessment of and Outlook for Nuclear Physics (NP2010 Committee) as part of the decadal studies of physics and astronomy conducted under the auspices of the Board on Physics and Astronomy. The principal goals of the study were to articulate the scientific rationale and objectives of the field and then to take a long-term strategic view of U.S. nuclear science in the global context for setting future directions for the field. The complete charge is presented in Appendix A.

The NP2010 Committee was composed of experts from universities and national laboratories in the United States, Canada, and Europe, mainly researchers in nuclear physics but also experts in other disciplines (see Appendix C for biographical information about committee members). The committee met four times in person, with the first meeting taking place on April 9-10, 2010, in Washington, D.C., and the fourth and final meeting on February 12-13, 2011 in Irvine, California. To provide an international context for research taking place in the United States, the NP2010 Committee heard from experts representing nuclear science from the Organisation for Economic Co-operation and Development global nuclear forum, from India, Europe, Canada, and Japan. The federal agencies that support nuclear physics research also briefed the committee, providing their perspectives on the issues to be addressed in this report. The committee thanks all those who met with it and supplied information. Their materials and discussions were valuable contributions to the committee's deliberations.

As chair and vice chair of the committee, we are particularly grateful to the committee members for their willingness to devote many hours to meeting and discussing all of the issues that arose and then to preparing the report. Finally, we thank the NRC staff for their guidance and assistance.

Stuart Freedman, *Chair*
Ani Aprahamian, *Vice Chair*
Committee on the Assessment of and Outlook for Nuclear Physics

# Acknowledgment of Reviewers

This report has been reviewed in draft form by individuals chosen for their diverse perspectives and technical expertise, in accordance with procedures approved by the National Research Council's (NRC's) Report Review Committee. The purpose of this independent review is to provide candid and critical comments that will assist the institution in making its published report as sound as possible and to ensure that the report meets institutional standards for objectivity, evidence, and responsiveness to the study charge. The review comments and draft manuscript remain confidential to protect the integrity of the deliberative process. We wish to thank the following individuals for their review of this report:

John Beacom, Ohio State University,
Paul Debevec, University of Illinois at Urbana-Champaign,
Gerry Garvey, Los Alamos National Laboratory,
Barbara Jacak, Stony Brook University,
Noemie Koller, Rutgers, The State University of New Jersey,
Alice Mignerey, University of Maryland,
Martin Savage, University of Washington,
Susan J. Seestrom, Los Alamos National Laboratory,
Brad Sherrill, Michigan State University, and
Priya Vashishta, University of Southern California.

Although the reviewers listed above have provided many constructive comments and suggestions, they were not asked to endorse the conclusions or recommendations, nor did they see the final draft of the report before its release. The review of this report was overseen by William H. Press, University of Texas at Austin, as monitor. Appointed by the NRC, he was responsible for making certain that an independent examination of this report was carried out in accordance with institutional procedures and that all review comments were carefully considered. Responsibility for the final content of this report rests entirely with the authoring committee and the institution.

# Contents

STUART JAY FREEDMAN
1944-2012

The committee dedicates this report to Stuart Freedman, its chair, who passed away unexpectedly on November 10, 2012. Stuart brought intellectual leadership, humor, friendship, and the highest standards of scientific excellence to his work. His loss is deeply felt throughout the community of nuclear physicists.

# Summary

This report provides a long-term assessment of and outlook for nuclear physics. The first phase of the report articulates the scientific rationale and objectives of the field, while the second phase provides a global context for the field and its long-term priorities and proposes a framework for progress through 2020 and beyond. The full statement of task for the committee is given in Appendix A.

Nuclear physics today is a diverse field, encompassing research that spans dimensions from a tiny fraction of the volume of the individual particles (neutrons and protons) in the atomic nucleus to the enormous scales of astrophysical objects in the cosmos. Its research objectives include the desire not only to better understand the nature of matter interacting at the nuclear level, but also to describe the nature of neutrinos and the state of the universe that existed at the big bang and that can now be studied in the most advanced colliding-beam accelerators, where strong forces are the dominant interactions.

The impact of nuclear physics extends well beyond furthering our scientific knowledge of the nucleus and nuclear properties. Nuclear science and its techniques, instruments, and tools are widely used to address major societal problems in medicine, border protection, national security, nonproliferation, nuclear forensics, energy technology, and climate research. Further, the tools developed by nuclear physicists often have important applications to other basic sciences—medicine, computational science, and materials research, among others—while its discoveries impact astrophysics, particle physics, and cosmology, and help to describe the physics of complex systems that arise in many fields.

In the second phase of the study, developing a framework for progress though

2020 and beyond, the committee carefully considered the balance between universities and government facilities in terms of research and workforce development and the role of international collaboration in leveraging future investments. The committee sought to address the means by which the balance between the various objectives of nuclear physics could be sustainable in the long term.

In summary, the committee finds that nuclear science in the United States is a vital enterprise that provides a steady stream of discoveries about the fundamental nature of subatomic matter that is enabling a new understanding of our world. The scientific results and technical developments of nuclear physics are also being used to enhance U.S. competition in innovation and economic growth and are having a tremendous interdisciplinary impact on other fields, such as astrophysics, biomedical physics, condensed matter physics, and fundamental particle physics. The application of this new knowledge is contributing in a fundamental way to the health and welfare of the nation. The committee's findings and recommendations are summarized below.

## FOLLOWING THROUGH WITH THE LONG-RANGE PLAN

The nuclear physics program in the United States has been especially well managed. Among the activities engaged in by the nuclear physics community is a recurring long-range planning process conducted under the auspices of the Nuclear Science Advisory Committee (NSAC) of the Department of Energy (DOE) and the National Science Foundation. This process includes a strong bottom-up emphasis and produces reports every 5 to 7 years that provide guidance to the funding agencies supporting the field. The choices made in NSAC's latest long-range plan, the Long Range Plan of 2007, have helped to move the field along and set it on its present course, and the scientific opportunities that process recognized as important will enable significant discoveries over the coming decade.

## Exploitation of Current Opportunities

Carrying through with the investments recommended in the 2007 Long Range Plan is the consequence of careful planning and sometimes difficult choices. The tradition of community engagement in the planning process has served the U.S. nuclear physics community well. A number of small and a few sizable resources have been developed since 2007 that are providing new opportunities to develop nuclear physics.

**Finding: By capitalizing on strategic investments, including the ongoing upgrade of the continuous electron beam accelerator facility (CEBAF) at the Thomas Jefferson National Accelerator Facility and the recently com-**

pleted upgrade of the Relativistic Heavy Ion Collider (RHIC) at Brookhaven National Laboratory, as well as other upgrades to the research infrastructure, nuclear physicists will confront new opportunities to make fundamental discoveries and lay the groundwork for new applications.

**Conclusion: Exploiting strategic investments should be an essential component of the U.S. nuclear science program in the coming decade.**

### The Facility for Rare Isotope Beams

After years of development and hard work involving a large segment of the U.S. nuclear physics community and the DOE, a major, world-leading new accelerator is being constructed in the United States.

**Finding: The Facility for Rare Isotope Beams is a major new strategic investment in nuclear science. It will have unique capabilities and will offer opportunities to answer fundamental questions about the inner workings of the atomic nucleus, the formation of the elements in our universe, and the evolution of the cosmos.**

**Recommendation: The Department of Energy's Office of Science, in conjunction with the state of Michigan and Michigan State University, should work toward the timely completion of the Facility for Rare Isotope Beams and the initiation of its physics program.**

### Underground Science in the United States

In recent decades the U.S. program in nuclear science has enabled important experimental discoveries such as the nature of neutrinos and the fundamental reactions fueling stars, often with the aid of carefully designed experiments conducted underground, where the backgrounds from cosmic radiation are especially low. The area of underground experimentation is a growing international enterprise in which U.S. nuclear scientists often play a key role.

**Recommendation: The Department of Energy, the National Science Foundation, and, where appropriate, other funding agencies should develop and implement a targeted program of underground science, including important experiments on whether neutrinos differ from antineutrinos, on the nature of dark matter, and on nuclear reactions of astrophysical importance. Such a**

program would be substantially enabled by the realization of a deep underground laboratory in the United States.

## BUILDING THE FOUNDATION FOR THE FUTURE

Nuclear physics in the United States is a diverse enterprise requiring the cooperation of many institutions. The subject of nuclear physics has evolved significantly since its beginnings in the early twentieth century. To continue to be healthy the enterprise will require that attention be paid to elements essential to the vitality of the field.

### Nuclear Physics at Universities

America's world-renowned universities are the discovery engines of the American scientific enterprise and are where the bright young minds of the next generation are recruited and trained. As with other sciences, it is imperative that the critical value-added role of universities and university research facilities in nuclear physics be sustained. Unfortunately, there has been a dramatic decrease in the number of university facilities dedicated to nuclear science research in the past decade, including fewer small accelerator facilities at universities as well as a reduction in technical infrastructure support for university-based research more generally. These developments could endanger U.S. nuclear science leadership in the medium and long term.

> **Finding: The dual role of universities—education and research—is important in all aspects of nuclear physics, including the operation of small, medium, and large facilities, as well as in the design and execution of large experiments at the national research laboratories. The vitality and sustainability of the U.S. nuclear physics program depend in an essential way on the intellectual environment and the workforce provided symbiotically by universities and the national laboratories. The fraction of the nuclear science budget reserved for facilities operations cannot continue to grow at the expense of the resources available to support research without serious damage to the overall nuclear science program.**

> **Conclusion: In order to ensure the long-term health of the field, it is critical to establish and maintain a balance between funding of operations at major facilities and the needs of university-based programs.**

A number of specific recommendations for programs to enhance the universities are discussed in the report. Many of these suggestions are not costly but could have significant impact. An example of a modest program that would enhance the recruitment of early career nuclear scientists and could be provided at relatively low cost is articulated in the following recommendation:

**Recommendation: The Department of Energy and the National Science Foundation should create and fund two national competitions: one a fellowship program for graduate students that would help recruit the best among the next generation into nuclear science and the other a fellowship program for postdoctoral researchers to provide the best young nuclear scientists with support, independence, and visibility.**

### Nuclear Physics and Exascale Computing

Enormous advances in computing power are taking place, and computers at the exascale are expected in the near future. This new capability is a game-changing event that will clearly impact many areas of science and engineering and will enable breakthroughs in key areas of nuclear physics. These include providing new understandings of, and predictive capabilities for, nuclear forces, nuclear structure and reaction dynamics, hadronic structure, phase transitions, matter under extreme conditions, stellar evolution and explosions, and accelerator science. It is essential for the future health of nuclear physics that there be a clear strategy for advancing computing capabilities in nuclear physics.

**Recommendation: A plan should be developed within the theoretical community and enabled by the appropriate sponsors that permits forefront computing resources to be exploited by nuclear science researchers and establishes the infrastructure and collaborations needed to take advantage of exascale capabilities as they become available.**

### Striving to Be Competitive and Innovative

Progress in science has always benefited from cooperation and from competition. For U.S. nuclear physics to flourish it must be competitive on the international scene, winning its share of the races to new discoveries and innovations. Providing a culture of innovation along with an understanding and acceptance of the appropriate associated risk must be the goal of the scientific research enterprise. The committee sees one particular aspect of science management in the United States where increased flexibility would have large and immediate benefits.

**Finding: The range of projects in nuclear physics is broad, and sophisticated new tools and protocols have been developed for successful management of**

the largest of them. At the smaller end of the scale, nimbleness is essential if the United States is to remain competitive and innovative on the rapidly expanding international nuclear physics scene.

**Recommendation: The sponsoring agencies should develop streamlined and flexible procedures that are tailored for initiating and managing smaller-scale nuclear science projects.**

### Prospects for an Electron-Ion Collider

Accelerators remain one of the key tools of nuclear physics, other fields of basic and applied research, and societal applications such as medicine. Modifying existing accelerators to incorporate new capabilities can be an effective way to advance the frontiers of the science. Of course it is the importance of the physics and of the potential discoveries enabled by the new capability that must justify the new investment. There is an initiative developing aimed at a new accelerator capability in the United States. Fortunately, the U.S. nuclear physics community has the mechanisms in place to properly evaluate this initiative. Currently there are suggestions that upgrades to either RHIC or CEBAF would enable the new capability.

**Finding: An upgrade to an existing accelerator facility that enables the colliding of nuclei and electrons at forefront energies would be unique for studying new aspects of quantum chromodynamics. In particular, such an upgrade would yield new information on the role of gluons in protons and nuclei. An electron-ion collider is currently under scrutiny as a possible future facility.**

**Recommendation: Investment in accelerator and detector research and development for an electron-ion collider should continue. The science opportunities and the requirements for such a facility should be carefully evaluated in the next Nuclear Science Long-Range Plan.**

Nuclear physics is a discovery-driven enterprise motivated by the desire to understand the fundamental mechanisms that account for the behavior of matter. Nevertheless, for its first hundred years, the new knowledge of the nuclear world has also directly benefited society through many innovative applications. As we move into the second century of nuclear physics the recommendations above will ensure a thriving and healthy field that continues to benefit society from new applications. Recently the stewardship of the nation's isotope program has been placed in the DOE Office of Nuclear Physics. This reorganization is appropriate and provides a fresh opportunity for the nuclear physics community to serve society by applying its sciences to the most important of today's problems in energy, health, and the environment. The isotopes program under the auspices of that office is expected to benefit rapidly from new innovations and developments. NSAC and its subcommittees have provided insightful reports that constitute a roadmap for the revitalized isotopes program. This advice is timely, coming when important decisions must be made. The committee sees these developments as an excellent example of how society's investments in nuclear physics can help resolve difficult challenges that face the nation.

# 1

# Overview

## INTRODUCTION

This fourth decadal assessment of nuclear physics by the National Research Council (NRC) comes exactly one century after Ernest Rutherford's discovery of the atomic nucleus. His visionary insight marked the beginning of nuclear physics. At 100 years, nuclear physics is a robust and vital science, with technological breakthroughs enabling experiments and computations that, in turn, are opening diverse new frontiers of exploration and discovery and addressing deep and important questions about the physical universe. Nuclear physicists today are advancing the frontiers of human knowledge in ways that are forcing us to revise our view of the cosmos, its beginnings, and the structure of matter within it. At the same time, these advances in nuclear physics are yielding applications that address some of the nation's challenges in security, health, energy, and education, as well as contributing innovations in technology and manufacturing that help drive our economy.

There have been stunning accomplishments and major discoveries in nuclear science since the last decadal assessment. Like Rutherford, today's nuclear scientists find that the data from well-crafted experiments often challenge them to revise their ideas about the structure of matter. Indeed, the matter that makes up all living organisms and ecosystems, planets and stars, throughout every galaxy in the universe, is made of atoms, and 99.9 percent of the mass of all the atoms in the universe comes from the nuclei at their centers, which are over 10,000 times smaller in diameter than the atoms themselves (the proton's radius is about a femtometer, or $10^{-15}$ m, a distance scale called the "femtoscale"). Although nuclei are incredibly small and dense, they are far from featureless: They are complex structures made of protons and neutrons, which themselves are complex structures made out of (as far as is known) elementary constituents known as quarks and gluons. Beyond what Rutherford could possibly have imagined, nuclear physics spans an enormous range of distance scales from well below the femtoscale upward to the scale of the universe itself.

The United States became a powerhouse in nuclear physics in the decades following the Manhattan Project. Today, vibrant nuclear physics programs are found, along with large and sophisticated nuclear physics laboratories, in most of the technologically advanced countries around the world. U.S. nuclear physicists often involve themselves in large collaborative efforts with scientists from many countries, carrying out experiments in the United States or abroad. Such efforts create new opportunities and optimize the deployment of the resources needed to germinate and sustain scientific progress and maintain intellectual leadership in nuclear physics. Managing these resources has become essential. To this end, the U.S. nuclear physics community has developed processes that build a community-wide vision, identifying which pathways will be the most effective and direct to scientific discoveries that open new vistas and drive the field. NRC's decadal assessments of nuclear physics have become one of the tools by which the field develops its roadmap. In this report, the Committee on the Assessment of and Outlook for Nuclear Physics ("the committee") assesses the state of nuclear physics at a time when it is rapidly evolving and new frontiers are opening up. It also looks at the prospects for the field in an international context.

Nuclear physics is broad and diverse in the questions it is answering and the challenges it faces on its many frontiers, as well as in its techniques and technologies. The committee frames this introduction with four overarching questions that span several of the traditional subfields of nuclear physics, that are central to the field as a whole, that reach out to other areas of science as well, and that together animate nuclear physics today:

(1) How did visible matter come into being and how does it evolve?
(2) How does subatomic matter organize itself and what phenomena emerge?
(3) Are the fundamental interactions that are basic to the structure of matter fully understood?
(4) How can the knowledge and technological progress provided by nuclear physics best be used to benefit society?

Accomplishments since the last NRC decadal assessment have brought us much closer to answering each of these four questions. In each case, recent research has revealed new physics discoveries and opened new frontiers for exploration. The questions are multifaceted, broad, and deep, and the challenges they pose provoke intriguing opportunities for the decade to come.

In the remainder of this introduction, these four questions are discussed in some detail and illustrated by a few vignettes. In Chapter 2 the scientific rationale and objectives of nuclear physics are articulated more fully. Chapter 2 is organized according to the main science areas within the field, but the four overarching questions cross the boundaries between these subfields, linking the discipline together as an intellectual whole while at the same time advancing on varied frontiers. Nuclear physics has Janus-like qualities, probing fundamental laws of nature that link it to particle physics while at the same time looking toward complex phenomena that emerge from the fundamental laws, as in atomic and condensed matter physics, and astrophysics and cosmology; zooming in on phenomena happening at the shortest distance scales that our best microscopes can see and zooming out to the stars and the cosmos. Because it sits in this liminal position between the fundamental and the emergent, between the microscopic and the astronomical, nuclear physics naturally addresses these central questions from varied angles, providing unique perspectives.

## HOW DID VISIBLE MATTER COME INTO BEING AND HOW DOES IT EVOLVE?

*The challenges posed by this question are shared by cosmologists, astronomers, particle physicists, and nuclear physicists alike. The universe is not entirely made of atoms and light: It also contains dark matter and dark energy—components that are known to exist because their gravitational influence on ordinary matter can be observed. But all the matter that can be seen—visible matter—is made of atoms, consisting of a tiny, compact nucleus and electrons orbiting around it. Atomic nuclei come in a very broad range of masses and electric charges. When the charges of the negative electronic cloud cancel out the positive charge of the nucleus, the atom is neutral. Interactions of the electronic clouds around nuclei enable the complex chemical processes that are essential for life and form the basis of our modern technological world. Atomic interactions are thus dictated by the atomic nuclei, as it is their charge that determines the electronic structure. Understanding nuclear physics and what goes on within the nuclei at the core of all visible matter starts with understanding the origins of the nuclei, light and heavy, and of the protons and neutrons of which they are made. How were the protons and neutrons created during the big bang? And, how did these protons and neutrons assemble into such a broad range of nuclei through nuclear transformations inside stars and stellar explosions? Nuclear science in concert with astrophysics attempts to answer these questions. The quest to understand how protons, neutrons, and nuclei form and evolve is fundamental to understanding our origins.*

One example of how nuclear physicists are learning how visible matter comes into being is provided by experiments at two accelerators: the Relativistic Heavy Ion Collider (RHIC) in Brookhaven, New York, and the Large Hadron Collider (LHC) in Geneva, Switzerland. By colliding nuclei at enormous energies, scientists are using these facilities to make little droplets of "big bang matter": the same stuff that filled the whole universe a few microseconds after the big bang. Using powerful detectors, they are seeking answers to questions about the properties of the matter that filled the microseconds-old universe that cannot be ascertained by any conceivable astronomical observations made with telescopes and satellites. Since the last decadal assessment of nuclear physics, research has shown that during the microsecond epoch, when the temperature of the universe was several trillion degrees, it was filled with a nearly perfect liquid that flowed with little viscous dissipation. This basic feature of big bang matter could only be discovered by recreating such matter in the laboratory.

As illustrated in Figure 1.1, sometime when the universe was about 10 microseconds old, this hot liquid cooled enough that it condensed, forming protons and neutrons (as well as other particles called pions), which, as far as is known, are the first complex structures ever created. These basic building blocks of all the visible

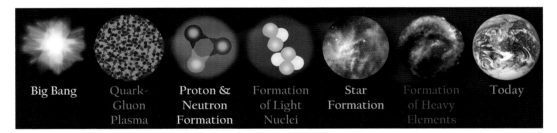

Big Bang    Quark-       Proton &     Formation    Star         Formation    Today
            Gluon        Neutron      of Light     Formation    of Heavy
            Plasma       Formation    Nuclei                    Elements

FIGURE 1.1 Nuclear physics in the universe. Over 99.9 percent of the mass of all the matter in all the living organisms, planets, and stars in all the galaxies throughout our universe comes from the nuclei found at the center of every atom. These nuclei are made of protons and neutrons that themselves formed a few microseconds after the big bang as the primordial liquid known as quark-gluon plasma cooled and condensed. The lightest nuclei (those at the centers of hydrogen, helium, and lithium atoms) formed minutes after the big bang. Other elements were formed later in nuclear reactions occurring deep within the early stars. Cataclysmic explosions of these early stars dispersed these heavy nuclei throughout the galaxy, so that as the solar system formed it contained nuclei of carbon, nitrogen, oxygen, silicon, iron, uranium, and many more elements, which ended up forming our planet and ourselves. SOURCE: Adapted from the Nuclear Science Wall Chart, developed by the Nuclear Science Division of the Lawrence Berkeley National Laboratory and the Contemporary Physics Education Project. Available at http://www.lbl.gov/abc/wallchart/index.html. Last accessed on May 30, 2012. Star Formation image: NASA/ESA and the Hubble Heritage Team (AURA/STScI/HEIC); Formation of Heavy Elements image: NASA/ESA/JHU/R. Sankrit and W. Blair; Today image: NASA.

matter in the universe today are under intense investigation at the Thomas Jefferson National Accelerator Facility (JLAB) in Newport News, Virginia. The facility there hosts an accelerator that can be thought of as an electron microscope so powerful that it can see inside protons and neutrons. Once the universe was a few minutes old, all the remaining neutrons in the universe paired up with protons to form light nuclei like those at the centers of helium and lithium atoms today; the remaining protons became the nuclei of hydrogen atoms. However, a panoply of elements exist in the world, not just hydrogen, helium, and lithium.

The processes of element synthesis go on today all across the universe, continuously creating new worlds. Nuclei are the fuel that powers the burning of stars and drives stellar explosions, some of which result in the formation of neutron stars, which can be thought of as nuclei of giant stellar masses. New nuclei, including those of which life is composed, are the ashes of stellar burning ejected into space by violent explosive events and stellar winds. The nuclear reactions that synthesize elements depend directly on the structure of the nuclei involved. This means that the element-by-element composition of matter in the universe today depends on features of thousands of nuclei, both the stable ones that are ordinarily seen and the unstable ones whose presence is fleeting. Many of these short-lived radioactive nuclei also play crucial roles in reactions taking place within the cores of nuclear reactors. Most important, very short-lived nuclei that are close to the limits on proton or neutron richness, beyond which no nuclei can exist, are thought to hold the secrets to the structure and formation of many of the stable nuclear species that surround us. There have been significant advances in the study of neutron-rich, proton-rich, and super heavy nuclei in the last decade, but the limits of nuclear existence still have not been demarcated. The characterization of nuclei near these limits that are so important to understanding the origins of visible matter also remains a challenge. Here, the Facility for Rare Isotope Beams (FRIB) at Michigan State University will utilize beams of short-lived nuclei to access the unknown regions of the nuclear landscape, providing new tools and new opportunities to address the challenge.

Significant advances in astronomy since the last decadal assessment have led to the discovery of very rare, very ancient stars whose composition reflects the production of elements by even earlier generations of stars, in some cases reaching back to stars formed from the debris of the very first generation of stellar explosions after the big bang. These ancient metal-poor stars are beginning to provide us with a chemical history of the galaxy, providing detailed information about the output of element-producing processes and in some cases hinting at previously unknown cycles of nuclear reactions responsible for making some of the elements heavier than iron. New facilities like FRIB will allow nuclear physicists to unravel the unknown properties of the nuclei and reactions that, in stars, are responsible for the creation of heavy elements.

Exploring the nuclear physics of the cosmos requires a broad range of experimental and theoretical approaches and can push nuclear science to its technical limits. Two important frontiers have arisen in the last decade and will be explored in the next decade with accelerators, detectors, and computers: (1) the fabrication and characterization in the laboratory of unstable nuclei that nature makes in stellar explosions and (2) the description of extremely slow nuclear reactions that are important for the understanding of stars, where they occur on astronomical timescales.

## HOW DOES SUBATOMIC MATTER ORGANIZE ITSELF AND WHAT PHENOMENA EMERGE?

*This question has been central to nuclear physics from Day One: Rutherford's 1911 discovery of the nuclei at the center of every atom framed it and provided the very first step toward answering it. Rutherford discovered heavy, apparently pointlike entities at the centers of atoms. He was correct to conclude that nuclei contain most of the mass of an atom, but little did he know how intricate their composition and structure would turn out to be. Nuclei are complex structures made of protons and neutrons. The number of protons in a nucleus determines the chemistry of the atom in which it is found—for example, all carbon nuclei have six protons, and this is what distinguishes carbon from oxygen, which possesses eight protons. As of today, nuclei containing as many as 118 protons have been found in nature or created in laboratories. The number of neutrons in a nucleus with a given number of protons can vary significantly. For example, although stable carbon nuclei contain either six or seven neutrons, short-lived variants have been discovered containing anywhere from 2 to 16 neutrons. There are far more isotopes (nuclei with a specified number of neutrons and protons) than elements (nuclei with a specified number of protons). Indeed, more than 3,100 different isotopes are known, and many thousands of additional isotopic species are believed to participate in element production in the stellar cauldrons of the cosmos. Understanding the patterns and regularities of their structure is one of the challenges of nuclear physics.*

*Remarkably, this challenge repeats itself at an even smaller length scale: Each proton and neutron is itself a complex structure made of (apparently) pointlike quarks, which are continually exchanging the force-carrying particles called gluons that provide the strong interactions binding the quarks into protons and neutrons (and pions and other short-lived complex structures). Unless, that is, one is talking about the matter that filled the microseconds-old universe, which was so hot that the matter that would later cool down and form protons, neutrons, and nuclei was a liquid of quarks and gluons. The complexities of the different structural elements of subatomic matter result in a plethora of possible states of matter at varying temperatures and densities. Understanding the structure of nuclei, and of their constituent protons and neutrons, as well as understanding the phases and phenomena that emerge when many of them get together, is among the grand challenges in nuclear physics. These challenges reso-nate across the many other areas of science in which macroscopic complexity emerges from large numbers of microscopic constituents obeying elementary rules. Some of the questions that arise are analogous to questions in other fields: How do large numbers of atoms organize themselves into materials: crystals, glasses, liquids, superfluids, and gases? How do large numbers of electrons arising from the atoms that make up these materials organize themselves to create metals, semiconductors, insulators, magnets, and superconductors? Just as the rich and varied forms of matter that make up the world originate in vast numbers of atoms and electrons interacting according to elementary microscopic laws, both theory and experiments have shown that large numbers of quarks or neutrons and protons or nuclei can also assemble themselves into a rich tapestry of possible phases of strongly interacting matter. The question of how many-body systems that are strongly correlated manifest new phases and new phenomena is a major intellectual thrust across many areas of physics. Examples of such bodies include novel superconductors, newly discovered topological patterns of quantum entanglement and quantum phase transitions in various condensed matter systems, warm dense plasmas, nuclear matter, quark-gluon plasma, and cold dense quark matter.*

One of the most exciting discoveries since the last decadal assessment is that the long-assumed periodicities in nuclear structure are, in fact, not always periodic. For about half a century, nuclei have been understood to be complex structures made of densely packed protons and neutrons with a structural organization that exhibits many regularities, analogous to the regularities in the structural organization of atoms that are manifest in the periodic table (see Figure 1.2). Recent experiments have shown that this need not always be so and have revealed that the familiar pattern of regularities occurs only for nuclei in which the numbers of protons and neutrons are not very different, as is the case for most known nuclei. For example, the number of neutrons it takes to "fill a shell"—the analogue of starting a new row in the periodic table, when structure starts to repeat itself—turns out to be different in short-lived nuclei with many more neutrons than protons than in stable nuclei with similar numbers of each. These recent discoveries challenge us to extend our understanding of the structure of matter and further motivate the study of very exotic nuclei—those that are extremely neutron-rich or extremely proton-rich or extremely heavy. These short-lived nuclei are but one example of the diverse patterns or phenomena that emerge as protons and neutrons organize themselves into nuclei.

In many instances, the quest to understand emergent phenomena connects nuclear physics directly with other areas of science in which interacting many-particle structures are central. For example, superconductivity in metals, in which pairs of electrons move in lockstep, and superfluidity in ultracold trapped atoms, in which pairs of atoms are created, both have analogues in nuclei (involving pairs of neutrons, pairs of protons, or possibly even proton-neutron pairs), in nuclear matter within neutron stars, and in dense quark matter that may exist at the very centers of neutron stars (where it is the quarks that form pairs). These are examples of collective quantum mechanical phenomena that can emerge only when many particles interact with one another. Many equally important emergent collective phenomena involving protons and neutrons in atomic nuclei will be studied at FRIB.

Remarkably, the basic story of having new phenomena emerge when elementary constituents organize themselves into complex structures repeats itself *within* single protons and neutrons. Although they are ordinarily thought of as the elementary constituents of nuclei, when protons and neutrons are looked at on shorter length scales, they themselves are revealed to be complex structures. Their

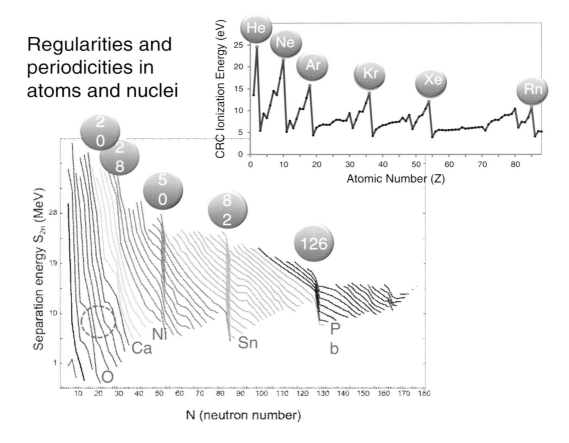

elementary constituents, quarks, are glued together by a force that is much stronger than the familiar electric and magnetic forces and that is associated with the exchange of elementary force-carrying particles called gluons. The experimental discoveries of quarks and gluons and the discovery of the laws that govern how they interact—called "quantum chromodynamics" (QCD)—are now more than 30 years old. And yet our understanding of how the properties of protons and neutrons arise from the interactions between their elementary constituents remains incomplete because the equations of QCD are simple to state but fiendishly difficult to solve. The underlying reason for this difficulty is that it is the interactions themselves that are the key feature. For example, while the electric and magnetic forces (mediated by massless photons) that bind an electron and a proton to form a hydrogen atom contribute only a tiny fraction of a percent to the mass of the atom, which is mostly just the mass of its constituents, it is now understood that approximately 99 percent of the mass of the protons and neutrons comes from

FIGURE 1.2 Regularities in the patterns of nuclei and of electrons in atoms. *Upper panel*: The elements in the periodic table are arranged in order of increasing atomic number, which is to say increasing numbers of electrons and protons per atom. (Atoms are electrically neutral; the number of protons in each atomic nucleus is balanced by the number of electrons orbiting the nucleus.) Elements having similar chemical properties and electronic structures appear in the same groups. This atomic periodicity, governed by the motion of the electrons in atoms, shows up in the behavior of the atomic ionization energy measured in electron volts of energy—namely, the energy needed to remove one electron from an atom. The chemical reactivity of an atom is determined by this ionization energy. Large jumps in the ionization energy occur in the noble gases (helium, neon, argon, krypton, xenon, and radon). High ionization energies mean that noble gases have very low chemical reactivity. *Lower panel*: Atomic nuclei themselves offer many examples of regularities and periodic behavior. The two-neutron separation energy (measured in millions of electron volts) is the energy required to remove a pair of neutrons from a nucleus that contains even numbers of protons and neutrons. This energy exhibits a sudden decrease immediately after specific "magic" neutron numbers (2, 8, 20, 28, 50, 82, 126). Nuclei with magic numbers of neutrons are more tightly bound than their neighbors with one extra neutron, making the former very much like noble gas atoms. (Different colors denote isotopes lying between proton magic numbers.) However, recent experiments have shown that the regular pattern of separation energies and other nuclear properties does not hold in sufficiently exotic nuclei, and that the magic numbers are "fragile." Examples that illustrate this phenomenon and are causing textbooks to be rewritten include the neutron-rich nuclei around magnesium-32, marked by a dashed circle, which do not seem to know about the magic neutron number $N = 20$, and the neutron-rich nuclei just to their right in the figure, which seem to almost ignore the magic neutron number $N = 28$. SOURCES: *(Upper right)* Data from David R. Lide (ed.), *CRC Handbook of Chemistry and Physics*, 84th Edition. Boca Raton, Florida: CRC Press. 2003, Section 10, Atomic, Molecular, and Optical Physics, Ionization Potentials of Atoms and Atomic Ions; (*Lower left*) E.J. Lingerfelt, M.S. Smith, H. Koura, Nuclear Masses Toolkit, 2012. Available at http://nuclearmasses.org.

the motion of the quarks inside them and from the mediators of the strong inter-action: massless gluons interacting with one another. The elementary masses of the quarks are so small that they contribute only a small fraction of the mass of the visible matter in the universe. So, the origin of 99 percent of the mass of the visible matter in the universe can be traced back to the energy of moving quarks and interacting gluons, according to Einstein's famous equation, $m = E/c^2$. The last decade has seen tremendous growth in the development of decisive experimental and theoretical tools that are for the first time giving us a precise look at the "shape" of protons and neutrons—for example, at the distribution of electric charge within them. Since quarks are the underlying charge carriers, such results are essential for understanding how the complex structure of protons and neutrons emerges from quarks and their QCD interactions.

One of the great surprises of the most recent decade has been the discovery that the elementary constituents within a proton or neutron have a significant net orbital motion, an "orbital angular momentum," as if the nucleons have hidden within them a circulating current of quarks and/or gluons. It has long been known that protons and neutrons have spin, a feature that makes medical diagnoses via magnetic resonance imaging possible. However, it was long assumed that this spin was due to the intrinsic spins of the elementary quarks lurking inside protons and neutrons rather than to their orbital motion. Just as the electron in a hydrogen atom has no orbital motion (no angular momentum) when it is in its lowest energy state, it was assumed that since protons and neutrons are the lowest energy states of three quarks, these quarks must have no orbital motion. Instead, experiments carried out during the last decade, discussed in more detail in Chapter 2 under "Momentum and Spin Within the Proton," have taught us that the spin of the proton and neu-tron appears to be largely due to orbital motion of the quarks or gluons trapped within them. This makes the internal structure of a single proton or neutron less like that of a hydrogen atom and more like that of a large nonspherical nucleus in which the collective orbital motion of hundreds of constituents is primarily responsible for the overall spin of the nucleus (see Figure 1.3). However, protons and neutrons are unique in having constituents within them that are moving at (ultrarelativistic) speeds very close to the speed of light. This discovery is motivat-ing a new generation of experiments at JLAB, Brookhaven National Laboratory, and many nuclear laboratories worldwide that will exploit advances in accelerator and detector technologies to fully characterize the distribution of mass and orbital motion within protons and neutrons.

As described above, the formation of the first protons and neutrons about 10 μsec after the big bang represented the earliest instance of the emergence of complex structures from the previously featureless primordial fluid. Although featureless in the sense that it was the same everywhere in the universe, the liquid of quarks and gluons (called the quark-gluon plasma, or QGP) that filled the

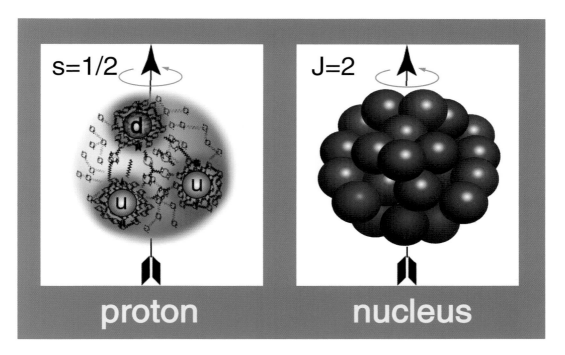

FIGURE 1.3 The spin of the proton is the sum of contributions from the spins and motions of all the quarks (u and d), quark-antiquark pairs (little circles), and gluons (connecting lines) within it. Recent experiments indicate that the sum of the orbital motion contributes more than the sum of all the spins, much as the total angular momentum of a large nonspherical nucleus is primarily the sum of contributions from the orbital motion of hundreds of protons (red) and neutrons (blue). SOURCE: (*left*) Lawrence S. Cardman, JLAB; (*right*) Witold Nazarewicz, University of Tennessee.

microseconds-old universe—and that is now being created and studied in experiments in which nuclei are collided at extreme energies—turns out itself to have very interesting properties that are emergent, in the sense that characteristics of the macroscopic fluid are far from apparent from the fundamental laws that govern the fluid's elementary constituents. As discussed in more detail in Chapter 2 under "Exploring Quark-Gluon Plasma," all observations of the droplets of QGP made in nuclear collisions over the last decade indicate that QGP acts more like a pureed soup—a liquid—than a dilute plasma in which particle-like quarks and gluons would be traveling appreciable distances between interactions, analogous to particle-like disturbances in ordinary gaseous atomic plasmas. Instead, liquid QGP responds to disturbances only with hydrodynamic waves, like those in water reacting to a dropped pebble. QGP is not the only known example of a fluid with collective properties but no apparent particle description: The challenge of understanding such liquids appears in several formerly disparate frontier areas of physics, including the study of ultracold atomic fluids; condensed matter systems such as oxide superconductors, which resist all conventional approaches to explaining their properties; and, perhaps most surprisingly, the fluid of quantum fluctuations found near black hole horizons. In the coming decade, nuclear physicists have the opportunity to understand how a complex liquidlike phase of matter emerges from the underlying elementary quarks and gluons, whose dynamics are well understood at very short distances. In addition to shedding light on the nature of the QGP that filled the microseconds-old universe, progress on this frontier could advance our understanding of phenomena that pose central challenges in many other areas of contemporary science.

We move into the twenty-first century with confidence that a full understanding of the fundamental theory of the strong interaction is within reach. QCD is a rich and enormously complex theory that describes complex structures, phases, and phenomena at the femtoscale. Applying QCD, and the effective nuclear theories that emerge from it at longer length scales, to develop a full understanding of the structure and properties of stars, nuclei, protons, and neutrons, and of the liquid QGP, will be one of the most compelling contributions of nuclear physics to science.

Within the next few years, a new generation of accelerators and detectors enabling new and perhaps unanticipated experimental discoveries, together with unprecedented computing power enabling groundbreaking calculations, will yield myriad new opportunities for advancing our understanding of the organization and properties of nuclear matter in all of its manifestations (see Figure 1.4). Current calculations are able to explain the basic properties of a proton or neutron, including its mass, in terms of interacting quarks and gluons. Now, such calculations are being extended to include more than one neutron or proton. Similarly, while the most advanced calculations done today that describe nuclei in terms of protons and neutrons and the empirical strong forces between them (without looking at the quarks and gluons inside the protons and neutrons) can explain the long lifetime of carbon-14 used for archaeological dating, a microscopic picture of the nuclear fission of uranium-235 still eludes us. Indeed, to fully explain the inner workings and multiscale complexity of protons, neutrons, and nuclei remains an enormous undertaking. The challenge is to include all the relevant physical features in deciphering truly complex problems rather than being forced to rely on simpler models that do not take into account the full physics involved. Examples of the pathways that have been mapped to overcome the daunting computational challenges include microscopic calculations of the properties of quark-gluon plasma and how it flows, how the quarks and gluons spin in a proton, and how protons and neutrons conspire to produce the collective phenomena and simple regularities seen in nuclei. A new generation of exascale computers, capable of performing a million trillion calculations per second, will allow simulations of nuclear fission, nuclear reactors, and hot and dense evolving environments such as those found in

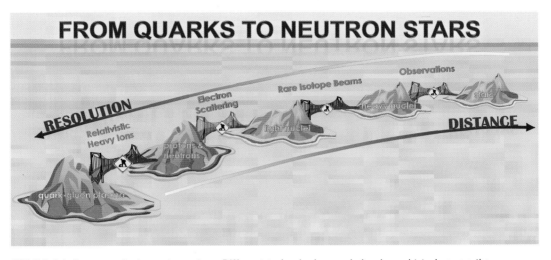

FIGURE 1.4 From quarks to neutron stars: Different technologies are being brought to bear on the myriad challenges in understanding nuclear matter at different spatial resolution scales or, equivalently, at different energy scales. At the shortest distance scales, relativistic heavy ion collisions are used to study quark-gluon plasma and how protons and neutrons and other hadrons condense from it as it cools. Electron-scattering experiments are used to study the complex structure of those protons and neutrons, with varying spatial resolution. Rare isotope beams are used to understand the patterns and phenomena that emerge as protons and neutrons form larger and larger nuclei. Nuclear phenomena occur on truly macroscopic distance scales in stars, in the nuclear reactions that drive certain classes of cataclysmic stellar explosions and in the description of the structure, formation, and cooling of neutron stars, which are basically gigantic nuclei. Building bridges of understanding between the physics at different spatial resolution scales is one of the paramount challenges facing contemporary nuclear science. For example, the most natural description of nuclei is in terms of neutrons and protons, and the most natural description of neutrons and protons is in terms of quarks and gluons. However, a rigorous connection between these two descriptive frameworks requires a description of the lightest nuclei in terms of quarks and gluons. This is the challenge for which the coming generation of accelerators, detectors, and computers is being designed, and is one of the great challenges for theoretical nuclear physics as well. SOURCE: Courtesy of Witold Nazarewicz, University of Tennessee.

inertial confinement fusion, nuclear weapons, and astrophysical phenomena and will provide a consistent picture of the fission data needed for national security and nuclear energy applications. Such computational capability, coupled with conceptual and algorithmic advances, will allow the physics of simple nuclei to be understood directly from QCD in terms of interacting quarks and gluons in a way that will serve as a benchmark for a rigorous computational approach to the full nuclear many-body problem. This bridge would link a century's worth of classic questions directly to the fundamental interactions that are now known to be basic to the structure of all matter.

## ARE THE FUNDAMENTAL INTERACTIONS THAT ARE BASIC
## TO THE STRUCTURE OF MATTER FULLY UNDERSTOOD?

*The first part of the answer to this question is known: The fundamental strong interactions between quarks and gluons (the laws of QCD) are known, and these elementary laws must be responsible first for the emergence of protons, neutrons, and their interactions, and then of nuclei. The interactions of QCD are not the only fundamental interactions we know of, however. All matter interacts by the gravitational force, and electrons are bound to nuclei (making atoms) by the electric and magnetic forces. Finally, the weak interactions ("weak" because they act only over distances much smaller than the size of a proton) are responsible for the radioactive decay of the majority of unstable nuclei—for example, carbon-14, used to estimate the age of carbon-bearing materials, and fluorine-18, used in medical imaging—and they determine the properties of the elusive neutrinos that pass through space, Earth, and our bodies, without us ever noticing. Nuclear scientists are able to utilize handpicked nuclei as laboratories in which to make extraordinarily precise measurements that provide stringent tests of our theories of all the fundamental interactions except gravity, which does, however, play a role in nuclear physics in the context of neutron stars, where neutron stars can be thought of as giant nuclei with a mass comparable to that of the sun. The theory that describes these fundamental interactions, namely QCD, together with the unified theory of electromagnetism and the weak interactions, is called the Standard Model. (It could more descriptively be called the "Theory of Visible Matter.") By testing the predictions of this theory for nuclear phenomena to exquisite precision, nuclear physicists are challenging the Standard Model and seeking evidence for new interactions that go beyond it.*

Nuclear physics has played a key role in the most significant revision to our understanding of the fundamental laws of nature that has come since the last decadal assessment—the discovery that neutrinos oscillate, transforming from one type, or "flavor," to another, even perhaps to a third type, and back and then repeating. The first evidence for this discovery came from Ray Davis's historic Nobel prize-winning experiment designed to measure the flux of neutrinos from the sun in conjunction with John Bahcall's precise modeling of how the sun shines, based on nuclear theory and nuclear data from laboratory experiments. Comparing the measured neutrino flux with solar model expectations, Davis found that about two-thirds of the expected neutrinos were missing, a mystery that remained unsolved for more than 30 years. Since the last decadal survey, however, two nuclear physics experiments, one at the Sudbury Neutrino Observatory in Canada (SNO) and the other in Japan (KamLAND), established convincingly that the neutrinos were not missing at all. Davis's experiment was sensitive to only one of the three flavors of neutrinos in nature, meaning that most of the neutrinos from the sun were hiding from Davis's detectors by oscillating into another flavor. Neutrino oscillations require that neutrinos must come with different masses, implying that at least two of them must have masses that are not zero. This discovery constitutes the first change in several decades in our understanding of the fundamental laws that govern the elementary constituents of all matter, namely the Standard Model (see Box 1.1). It opens new questions, the most profound of which are the determination of the average neutrino mass and the source of thier mass and the determination of whether neutrinos are their own antiparticles. Concerted efforts to answer these and other questions are now being mounted by nuclear physicists in a mutually beneficial partnership with their particle physics colleagues.

Physicists do not expect the appearance of neutrino masses to be the last word in the quest to understand the laws of nature at the level of elementary particles and their interactions. Our current understanding, as codified in the Standard Model, has had an extraordinary run of success in describing many phenomena, but it is incomplete. Nuclear and particle physicists are seeking a new Standard Model (NSM), which will incorporate the many successes of the Standard Model but will in addition provide an understanding of aspects of physics that are now mysterious. Questions here include: Why do quarks and electrons have the masses that they have? What is the nature of the dark matter and dark energy that pervade the universe? and Why is the universe filled with matter but little antimatter? One approach to answering these questions, led by particle physicists, is to push back the high-energy frontier, seeking to create whatever new particles and new interactions that may exist in the NSM in proton-proton collisions at the Large Hadron Collider. An alternative approach, where nuclear physicists play a leading role, is to make advances on the precision frontier, where exquisitely sensitive measurements may reveal tiny deviations from Standard Model predictions and point to the fundamental symmetries of the NSM. For example, the symmetries of the Standard Model do allow a neutron to have a very tiny permanent separation between the center of mass of the positively charged quarks and the center of mass of the negatively charged quarks within it, but many ideas for the NSM allow for a possibly larger charge separation, known as the neutron dipole moment. In the coming decade, nuclear physicists are planning a campaign to detect such an effect or at least greatly reduce the experimental limits on it. The detection of a charge separation larger than that allowed by the Standard Model could have decisive implications for our understanding of the NSM and would naturally accommodate mechanisms for the generation of an excess of matter over antimatter when the universe was a trillionth of a microsecond or less old.

---

**Box 1.1**
**The Fundamental Matter Particles of the Standard Model,**
**Also Sometimes Called the "Theory of Visible Matter"**

There are six quarks in the Standard Model: the up (u), down (d), charm (c), strange (s), top (t), and bottom (b) quarks. Quarks are matter particles that emit and absorb massless gluons, meaning that they experience the strong interactions. The matter particles that do not participate in the strong interactions (called "leptons") include the electron (e) and its two cousins, the muon ($\mu$) and the tauon ($\tau$). Leptons and the quarks are charged, meaning that they emit and absorb massless photons and thus experience electric and magnetic interactions. Neutrinos ($\nu$) are the only known fundamental matter particles that do not absorb or emit either photons or gluons. All matter particles, including neutrinos, emit and absorb W and Z bosons; the consequent interactions are weak because the W and Z bosons are heavy, each having about half the mass of the heaviest known matter particle, the top quark. All matter particles also feel the force of gravity. All the particles in Figure 1.1.1 except the neutrino have antiparticles with the same mass and the opposite electric charge. Each of the three neutrinos may have an antiparticle or may be its own antiparticle; ongoing nuclear physics experiments aim to determine which.

The three neutrinos have different masses and so, when labeled by their masses as in this figure, they can be called "light," "medium," and "heavy." The pattern of neutrino masses shown is one of the possibilities suggested by the recent discovery of neutrino oscillations, captured in the pie chart for each neutrino. The three colors reflect the flavors of the electron, muon, and tauon charged leptons. They show that the only way to construct a neutrino that is the exact partner of the electron (called an "electron neutrino," it is blue in this diagram) is to combine neutrinos with differing masses in a certain way. Nuclear reactions in the sun produce electron neutrinos. And, all Standard Model processes in which a neutrino is made produce the exact partner of one of the charged leptons. The fact that these are combinations of neutrinos with differing masses is what causes the neutrino to oscillate as it flies through space. (Quarks also show this mixing property but to a much lesser degree.) Although the discovery of neutrino oscillations has given us good information about the differences between the three neutrino masses, the mass of the lightest neutrino is not known precisely; it could even be zero. Ongoing nuclear physics experiments seeking to measure the average neutrino mass directly, not via oscillations, will help. But we do not yet have any fundamental understanding of the pattern of the masses of the 12 Standard Model matter particles, in particular of why the neutrinos are millions of times lighter than any of the other particles.

## HOW CAN THE KNOWLEDGE AND TECHNOLOGICAL PROGRESS PROVIDED BY NUCLEAR PHYSICS BEST BE USED TO BENEFIT SOCIETY?

*Nuclear physics is not only a basic scientific enterprise, but it also has stunning practical applications. These in themselves can justify the cost and effort of the research, going beyond the basic knowledge that has been gained. Nuclear science has a many-decade-long history of accomplishments that benefit our health, the economy, and our safety and security: The societal benefits derived from nuclear physics are by now ubiquitous. This track record continues, with many new accomplishments since the last decadal assessment and many more under development. Smoke detectors in our*

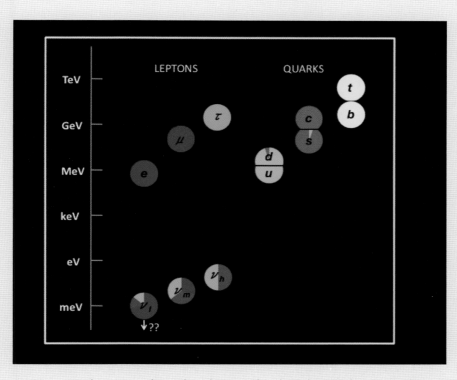

FIGURE 1.1.1 The masses of particles. The vertical scale is the particle mass in electron-volts, with each tick representing a 1,000-fold increase. SOURCE: Courtesy of R.G. Hamish Robertson, University of Washington.

homes, new medical diagnostic imaging methods, therapies using ion beams and new isotopes for cancer treatments, and new methods for assessing breaches in national and homeland security are just some of the ways that nuclear physics makes a difference to our safety, health, and security. Technological advances driven by advances in nuclear physics, which range from particle accelerators (most of which are now used either for medical purposes or in the semiconductor industry) to supercomputers, make significant contributions to our economy. A mutually beneficial synergy has developed in which a fundamental intellectual enterprise has consistently produced technological gains that, pursued with societal benefit in mind, have more than compensated the *public support required to pursue this science. Also beneficial to society is informing people about the discoveries from nuclear physics and explaining to them the origin and structure of matter and the fundamental interactions.*

Positron emission tomography exemplifies the synergy between nuclear science, technological advances, and benefits to society. This medical imaging technique has become a powerful new tool for the diagnosis of cancer. The positron sources as well as the highly segmented crystalline detector elements come directly from nuclear physics research. Another example of synergy: Over the last two decades, nuclear physicists interested in the structure of the neutron have developed spin-polarized helium-3 targets, and it now turns out that these very techniques can be used to make spin-polarized helium-3 or xenon-129. These, in turn, can be introduced into the air a patient breathes, allowing for a new kind of magnetic resonance imaging (MRI) of the lungs. Without these developments, MRI could not be used to image gases and thus would not be able to accurately visualize lung function.

Nuclear science plays a role in treatment as well as in diagnosis. Nuclear medicine is a well-established field within medical research and therapy, with techniques that originated in nuclear science now used as a matter of course in the irradiation of tumors with high-energy particles. One of the many exciting advances in treatment being pursued today is targeted radionuclide therapy, which has been the most highly sought-after goal of nuclear medicine physicians and scientists for decades. Targeted therapy involves attaching a "targeting molecule" to a relatively short-lived radioactive isotope. The isotope emits radiation (alpha particles, for example) that reacts strongly with nuclei of atoms that comprise the tissues in the body and so deposits most of its energy nearby. The biologically active targeting molecules are carefully designed to bind to receptors on cancer tumor cells. When the radioactive nuclei attached to the targeting molecules bound to the tumor decay, they deliver a lethal dose of radiation only to the tumor tissue. By careful construction, the targeting molecule will pass through the body quickly if it does not bind to tumor cells, thus minimizing the exposure of healthy tissue to radiation. The use of these techniques in human clinical trials and in actual clinical therapy has just started. Two radiopharmaceuticals are now in use to treat non-Hodgkins lymphoma. And, recently, researchers at the Institute for Transuranium Elements in Karlsruhe, Germany, treated neuroendocrine tumors with the alpha-emitting bismuth-213 nucleus attached to a biological molecule that targets these particular tumor cells; they found in a small initial trial with human patients a reduction in the size of some tumors with no discernible negative side effects. If this approach can become routine, the treatment of cancer will undergo a paradigm shift. Nuclear scientists play an essential role in this interdisciplinary effort, which blends biology, medicine, modern technology, and nuclear physics.

The years since 9/11 have seen important advances in nuclear forensics. An attack using a nuclear or radiological explosive device would of course be catastrophic, but it would also raise a set of urgent and crucial questions: What was exploded? Who did it? Do they have more? Was the device improvised or sophisticated? Did they steal it, and if so from where? Is the material reactor-grade or weapons-grade fuel? How old is it? Nuclear forensics refers to the techniques and capabilities needed to answer these questions. It can be likened to forensics-style exercises in nuclear astrophysics, whereby scientists analyze and evaluate the debris left behind by a stellar-scale nuclear explosion. Both efforts—nuclear astrophysics and nuclear forensics—are led by nuclear physicists.

The last decade has also seen major advances in the use of nuclear physics techniques to detect heavy nuclei like uranium or plutonium in a truck or cargo container—one such technique might detect the cosmic ray muons, elementary particles similar to electrons, that scatter off these elements at large angles. The techniques are based on well-understood basic nuclear physics, reminiscent of Rutherford's early experiments, but their application to the detection of nuclear contraband crossing U.S. borders is new. The detector and computational challenges are related to very recent developments in basic nuclear physics.

Nuclear physics has long been a driver in the development of accelerators and computers, both of which are prevalent in our lives and in many sectors of the economy. Solving the design challenges associated with the building of very high energy accelerators being used to probe the fundamental nature of the matter in our universe will bring advances that improve the more than 30,000 accelerators used around the world for radiotherapy, for ion implantation to precisely embed dopants in semiconductor chips, and for other applications from developing new materials to improving food safety and benefiting other areas of industrial and biomedical research. Advancing nuclear science also drives innovations in computer architecture. For example, when IBM developed the Blue Gene line of computers that have become successful commercial machines with an impact on climate science, genomics, protein folding, materials science, and brain simulation, it employed a paradigm that had been developed first for lattice QCD machines—in fact, IBM employed people who had previously designed a computer called the QCDOC (QCD on a chip).

These examples all show how investment in nuclear physics has benefits beyond addressing the fundamental overarching questions earlier in this chapter. These investments are yielding progress on some of the nation's biggest challenges as well as innovations that help to drive the economy.

## PLANNING FOR THE FUTURE

As the complexity of the main challenges in the field has grown, so have the cost and size of the experimental nuclear physics tools. What began 100 years ago primarily as efforts of individuals or small groups has grown into a mix of small and large groups working as teams, both here and abroad. The U.S. nuclear physics community has developed a number of complementary processes for establishing consensus and setting priorities and future directions. The Division of Nuclear Physics in the American Physical Society, one of the most active divisions, provides help with planning and outreach for the benefit of nuclear physics. Another effective element is the Long-Range Planning process organized by the Nuclear Science Advisory Committee (NSAC) of the Department of Energy and the National Science Foundation. Using this tool, the community has been establishing its priorities and providing guidance and advice to the funding agencies. The present decadal assessment of nuclear physics brings experts from the diverse areas of the field to assess the achievements and provide a forward-looking vision of the new horizon. Anticipating needs for personnel and for building new facilities as well as developing and improving infrastructure for the field are all important components of the planning process. The charge for this study reflects the mission of decadal studies:

> The new 2010 NRC decadal report will prepare an assessment and outlook for nuclear physics research in the United States in the international context. The first phase of the study will focus on developing a clear and compelling articulation of the scientific rationale and objectives of nuclear physics. This phase would build on the 2007 NSAC Long-range Plan Report, placing the near-term goals of that report in a broader national context.
>
> The second phase will put the long-term priorities for the field (in terms of major facilities, research infrastructure, and scientific manpower) into a global context and develop a strategy that can serve as a framework for progress in U.S. nuclear physics through 2020 and beyond. It will discuss opportunities to optimize the partnership between major facilities and the universities in areas such as research productivity and the recruitment of young researchers. It will address the role of international collaboration in leveraging future U.S. investments in nuclear science. The strategy will address means to balance the various objectives of the field in a sustainable manner over the long term.

This present report offers the committee's assessment and outlook. Chapter 2 summarizes the main scientific areas and the science questions addressed by nuclear physics, focusing on accomplishments since the last decadal assessment and directions for the decade to come. From the beginning the diversity of the science is evident in the range of topics, from the behavior of quarks and gluons to the universe. In this introduction, the committee has highlighted the interconnections of these main scientific areas with each other.

In Chapter 3 as well as elsewhere in this report, some of the ways in which society benefits from applications of nuclear physics are emphasized, and snapshots of various important uses of the knowledge and know-how gained from nuclear physics are provided. Again, the topics of application are astonishingly diverse.

Nuclear physics in the global context is described in Chapter 4. Resulting from remarkably productive international cooperation, the global program in nuclear physics combines competition, cooperation, and communication in a way that is benefiting all the participants and accelerating scientific progress.

Chapter 5 addresses the important issues related to decision-making processes. The critical NSAC Long Range Planning exercise and other less structured global planning processes have become vital for keeping the nuclear physics enterprise on a successful path to the future. Workforce issues are explored in this chapter along with the steps being taken to ensure a productive workforce in the coming years. Finally, the committee discusses its findings and recommendations in Chapter 6.

# 2

# Science Questions

**INTRODUCTION**

This chapter discusses in more detail the recent accomplishments and directions that are expected to be taken in nuclear physics in upcoming years. Where the discussion in Chapter 1 focused on four overarching questions being addressed by the field, this chapter is separated into more traditional subfields of nuclear physics—(1) nuclear structure, whose goal is to build a coherent framework for explaining all properties of nuclei and nuclear matter and how they interact; (2) nuclear astrophysics, which explores those events and objects in the universe shaped by nuclear reactions; (3) quark-gluon plasma, which examines the state of "melted" nuclei and with that knowledge seeks to shed light on the beginnings of the universe and the nature of those quarks and gluons that are the constituent particles of nuclei; (4) hadron structure, which explores the remarkable characteristics of the strong force and the various mechanisms by which the quarks and gluons interact and result in the properties of the protons and neutrons that make up nuclei; and (5) fundamental symmetries, those areas on the edge of nuclear physics where the understandings and tools of nuclear physicists are being used to unravel limitations of the Standard Model and to provide some of the understandings upon which a new, more comprehensive Standard Model will be built.

## PERSPECTIVES ON THE STRUCTURE OF ATOMIC NUCLEI

The goal of nuclear structure research is to build a coherent framework that explains all the properties of nuclei, nuclear matter, and nuclear reactions. While extremely ambitious, this goal is no longer a dream. With the advent of new generations of exotic beam facilities, which will greatly expand the variety and intensity of rare isotopes available, new theoretical concepts, and the extreme-scale computing platforms that enable cutting-edge calculations of nuclear properties, nuclear structure physics is poised at the threshold of its most dramatic expansion of opportunities in decades.

The overarching questions guiding nuclear structure research have been expressed as two general and complementary perspectives: a microscopic view focusing on the motion of individual nucleons and their mutual interactions, and a mesoscopic one that focuses on a highly organized complex system exhibiting special symmetries, regularities, and collective behavior. Through those two perspectives, research in nuclear structure in the next decade will seek answers to a number of open questions:

- What are the limits of nuclear existence and how do nuclei at those limits live and die?
- What do regular patterns in the behavior of nuclei divulge about the nature of nuclear forces and the mechanism of nuclear binding?
- What is the nature of extended nucleonic matter?
- How can nuclear structure and reactions be described in a unified way?

New facilities and tools will help to explore the vast nuclear landscape and identify the missing ingredients in our understanding of the nucleus. A huge number of new nuclei are now available—proton rich, neutron rich, the heaviest elements, and the long chains of isotopes for many elements. Together, they comprise a vast pool from which key isotopes—designer nuclei—can be chosen because they isolate or amplify specific physics or are important for applications.

At the same time, research with intense beams of stable nuclei continues to produce innovative science, and, in the long term, discoveries at exotic beam facilities will raise new questions whose answers are accessible with stable nuclei.

Examples of the current program that offer a glimpse into future areas of inquiry are the investigation of new forms of nuclear matter such as neutron skins occurring on the surfaces of nuclei having large excesses of neutrons over protons, the ability to fabricate the superheavy elements that are predicted to exhibit unusual stability in spite of huge electrostatic repulsion, and structural studies in exotic isotopes whose properties defy current textbook paradigms.

Hand in hand with experimental developments, a qualitative change is taking place in theoretical nuclear structure physics. With the development of new concepts, the exploitation of symbiotic collaborations with scientists in diverse fields, and advances in computing technology and numerical algorithms, theorists are progressing toward understanding the nucleus in a comprehensive and unified way.

## Revising the Paradigms of Nuclear Structure

### Shell Structure: A Moving Target

The concept of nucleons moving in orbits within the nucleus under the influence of a common force gives rise to the ideas of shell structure and resulting magic numbers. Like an electron's motion in an atom, nucleonic orbits bunch together in energy, forming shells, and nuclei having filled nucleonic shells (nuclear "noble gases") are exceptionally well bound. The numbers of nucleons needed to fill each successive shell are called the magic numbers: The traditional ones are 2, 8, 20, 28, 50, 82, and 126 (some of these are exemplified in Figure 2.1). Thus a nucleus such as lead-208, with 82 protons and 126 neutrons, is doubly "magic." The concept of magic numbers in turn introduces the idea of valence nucleons—those beyond a magic number. Thus, in considering the structure of nuclei like lead-210, one can, to some approximation, consider only the last two valence neutrons rather than all 210. When proposed in the late 1940s, this was a revolutionary concept: How could individual nucleons, which fill most of the nuclear volume, orbit so freely without generating an absolute chaos of collisions? Of course, the Pauli exclusion principle is now understood to play a key role here, and the resulting model of nucleonic orbits has become the template used for over half a century to view nuclear structure.

One experimental hallmark of nuclear structure is the behavior of the first excited state with angular momentum 2 and positive parity in even-even nuclei. This state, usually the lowest energy excitation in such nuclei, is a bellwether of structure. Its excitation energy takes on high values at magic numbers and low values as the number of valence nucleons increases and collective behavior emerges. The picture of nuclear shells leads to the beautiful regularities and simple repeated patterns, illustrated in Figure 1.2 and seen here in the energies of the $2^+$ states shown at the top of Figure 2.2. The concept of magic numbers was forged from data based on stable or near-stable nuclei. Recently, however, the traditional magic numbers underwent major revisions as previously unavailable species became accessible. The shell structure known from stable nuclei is no longer viewed as an immutable construct but instead is seen as an evolving moving target. Indeed the elucidation of changing shell structure is one of the triumphs of recent experiments in nuclear structure at exotic beam facilities worldwide. For example, experiments

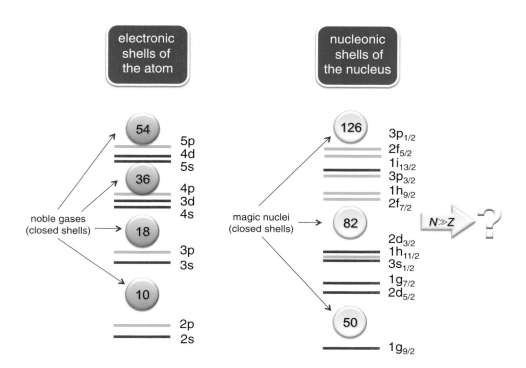

FIGURE 2.1 Shell structure in atoms and nuclei. *Left:* Electron energy levels forming the atomic shell structure. In the noble gases, shells of valence electrons are completely filled. *Right:* Representative nuclear shell structure characteristic of stable or long-lived nuclei close to the valley of stability. In the "magic" nuclei with proton or neutron numbers 2, 8, 20, 28, 50, 82, and 126, which are analogous to noble gases, proton and/or neutron shells are completely filled. The shell structure in very neutron-rich nuclei is not known. New data on light nuclei with N >> Z tell us that significant modifications are expected. SOURCE: Adapted and reprinted with permission from K. Jones and W. Nazarewicz, 2010, *The Physics Teacher* 48 (381). Copyright 2010, American Association of Physics Teachers.

at Michigan State University (MSU) in the United States and at the Gesellschaft für Schwerionenforschung (GSI) have shown that in the very neutron-rich isotope oxygen-24, with 8 protons and twice as many neutrons, N = 16 is, in fact, a new magic number.

One of the most interesting regions exhibiting the fragility of magic numbers is nuclei with 12 to 20 protons and 18 to 30 neutrons. The experimental evidence is exemplified in the lower portion of Figure 2.2 by the energies of the first excited $2^+$ states in this region. The figure shows the disappearance of neutron number N = 20 as a magic number in magnesium while it persists for neighboring elements.

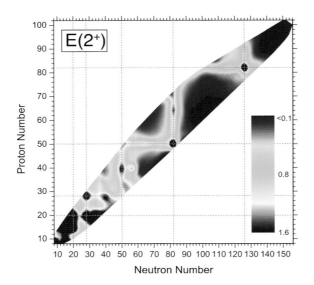

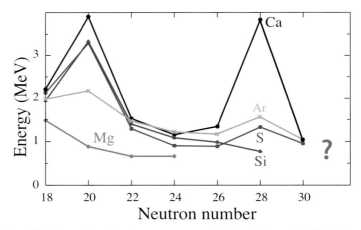

FIGURE 2.2 Measured energies of the lowest 2+ states in even-even nuclei. *Top:* Color-coded Z-N plot spanning the entire nuclear chart clearly show the filaments of magic behavior at particular neutron and proton numbers (denoted by dashed lines) and the lowering of these states as nucleons are added and collective behavior emerges. The legend bar relates the colors to an energy scale in MeV. *Bottom:* Close-up view of the data for the neutron-rich magnesium (Mg), silicon (Si), sulfur (S), argon (Ar), and calcium (Ca) isotopes. The fingerprint of magic numbers is missing in the neutron-rich isotopes of Mg, Si, S, and Ar, in which the "standard" magic numbers at either N = 20 or 28 have dissipated. As of 2011, no data exist on Si, S, and Ar nuclei with N = 32. SOURCES: *(Top)* Courtesy of R. Burcu Cakirli, Max Planck Institute for Nuclear Physics, private communication, 2011; *(bottom)* Courtesy of Alexandra Gade, MSU.

Similarly, N = 28 loses its magic character for silicon, sulfur, and argon, while calcium, which is also magic in protons, retains its doubly magic character at N = 28.

There are at least three factors leading to such changes in shell structure: changes in how nucleons interact with each other as the proton-neutron asymmetry varies, the influence of scattering and decay states near the isotopic limits of nuclear existence (the "drip lines"), and the increasing role of many-body effects in weakly bound nuclei where correlations determine the mere existence of the nucleus. This new perspective on shell structure affects many facets of nuclear structure, from the existence of short-lived light nuclei, to the emergence of collectivity, to the stability of the superheavy elements.

Recent studies of calcium, nickel, and tin isotopes using techniques such as Coulomb excitation and light-ion single nucleon transfer reactions, both near traditional magic numbers and along extended isotopic chains, are beginning to answer questions about effective internucleon forces in the presence of large neutron excess, the relevance of the detailed shell-model template in the presence of weak binding, and the nature of nuclear collective motion. Excellent tests of the nuclear shell model were offered by recent studies of the tin (Sn) isotopes. The nucleus tin has a magic number (50) of protons, and its short-lived isotopes tin-100 and tin-132, with 50 and 82 neutrons, respectively, are expected to be rare examples of new doubly magic heavy nuclei. Unique data in the tin-132 region (see Figure 2.3) shows that tin-132 indeed behaves as a good doubly magic nucleus. Other experiments providing data around tin-100, in particular the first structural information on tin-101, have led to theoretical surprises. Further tests of shell structure and interactions in the heaviest elements will be discussed below.

It is expected that the shell model will undergo sensitive tests in the region of superheavy nuclei, whose very existence hinges on a dynamical competition between short-range nuclear attraction and huge long-range Coulomb repulsion. Interestingly, a similar interplay takes place in low-density, neutron-rich matter found in crusts of neutron stars, where "Coulomb frustration" produces rich and complex collective structures, discussed later in this chapter in "Nuclear Astrophysics." Figure 2.4 shows the calculated shell energy—that is, the quantum enhancement in nuclear binding due to the presence of nucleonic shells. The nuclei from the tin region are excellent examples of the shell-model paradigm: the magic nuclei with Z = 50, N = 50, and N = 82 have the largest shell energies, and the associated closed shells provide exceptional stability. In superheavy nuclei, the density of single-particle energy levels is fairly large, so small energy shifts, such as the regions of enhanced shell stabilization in the super-heavy region near N = 184, are generally expected to be fairly broad; that is, the notion of magic numbers and the energy gaps associated with them becomes fluid there.

Another dimension in studies of shells in nuclei has been opened by precision studies, at the Thomas Jefferson National Accelerator Facility (JLAB) and at the

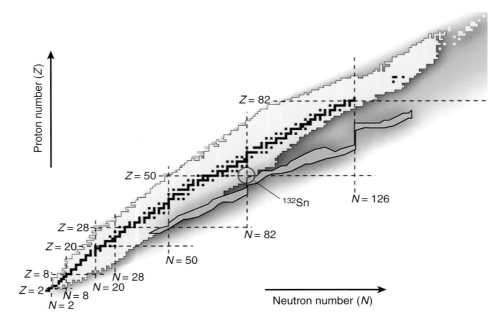

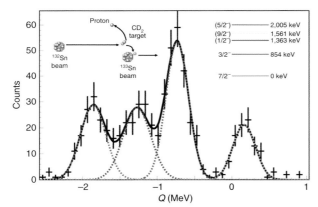

FIGURE 2.3 *Top:* All known nuclides are shown as black (if stable) or yellow (unstable). Dashed lines indicate the traditional magic numbers of protons and neutrons. Two doubly-magic nuclei, tin-132 and nickel-78, are adjacent to the r-process region (blue) of as-yet-unseen nuclides that are thought to be involved in the creation of the heaviest elements in supernovae. By adding neutrons or protons to a stable nucleus, one enters the territory of radioactive nuclei, first long-lived, then short-lived, until finally the nuclear drip line is reached, where there is no longer enough binding force to prevent the last nucleons from dripping off the nuclei. The proton and neutron drip lines form the borders of nuclear existence. *Bottom:* Experimental spectrum for a transfer reaction in which an incident deuteron grazes a tin-132 target, depositing a neutron to make tin-133 with detection of the exiting proton (that is, $d + \text{tin-132} \rightarrow p + \text{tin-133}$). The solid line shows a fit to the four peaks shown in green, red, blue, and lavender in the level scheme (inset). The top left inset displays a cartoon of the reaction employed. The investigations revealed that low energy states in tin-133 have even purer single-particle character than their counterparts in lead-209, outside the doubly-magic nucleus lead-208, the previous benchmark. SOURCE: *(Top)* Reprinted by permission from Macmillan Publishers Ltd., B. Schwarzschild. August 2010. *Physics Today* 63:16, copyright 2010; *(Bottom)* Reprinted by permission from Macmillan Publishers Ltd., K.L. Jones, A.S. Adekola, D.W. Bardayan, et al. 2010. *Nature* 465: 454, copyright 2010. Portions of the figure caption are extracted from K.L. Jones, W. Nazarewicz, 2010. Designer nuclei – making atoms that barely exist, *The Physics Teacher* 48: 381.

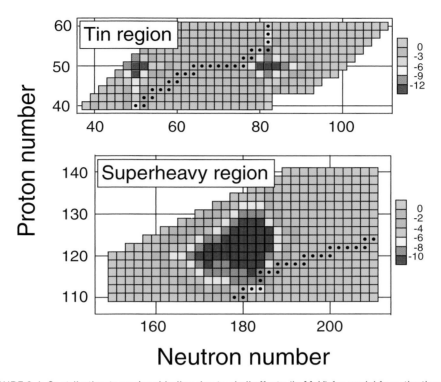

FIGURE 2.4 Contribution to nuclear binding due to shell effects (in MeV) for nuclei from the tin region *(top)* and super-heavy elements *(bottom)* calculated in the nuclear density functional theory. The nuclei colored in darker red are those whose binding is most enhanced by quantum effects. The nuclei predicted to be stable to beta decay are marked by dots. SOURCE: Reprinted and adapted from M. Bender, W. Nazarewicz, and P.G. Reinhard, Shell stabilization of super- and hyperheavy nuclei without magic gaps, *Physics Letters B* 515: 42, Copyright 2001, with permission from Elsevier.

Japanese National Laboratory for High Energy Physics (KEK), of hypernuclei—nuclei that contain at least one hyperon, a strange baryon, in addition to nucleons. By adding a hyperon, nuclear physicists can explore inner regions of nuclei that are impossible to study with protons and neutrons, which must obey the constraints imposed by the Pauli principle. The experimental work goes hand in hand with advanced theoretical calculations of hyperon-nucleon and hyperon-hyperon interactions, with the ultimate goal being the comprehensive understanding of all baryon-baryon interactions.

*Exploring and Understanding the Limits of Nuclear Existence*

An important challenge is to delineate the proton and neutron drip lines—the limits of proton and neutron numbers at which nuclei are no longer bound by the strong force and nuclear existence ends—as far into the nuclear chart as possible (see Figure 2.3 [top]). For example, experiments at MSU have produced the heaviest magnesium and aluminum isotopes accessible to date and have shown that magnesium-40, aluminum-42, and possibly aluminum-43 exist. Nuclei near the drip lines are very weakly bound quantum systems, often with extremely large spatial sizes. In recent years, experiments at Argonne National Laboratory (ANL), TRIUMF, Grand Accélérateur National d'Ions Lourds (GANIL), GSI, the European Organization for Nuclear Research (CERN), and Rikagaku Kenky jo (RIKEN) using high-precision laser spectroscopy have determined the charge radii of halo nuclei helium-6, helium-8, beryllium-11, and lithium-11 with an accuracy of 1 percent through the determination of isotope shifts of atomic electronic levels. With the advanced-generation Facility for Rare Isotope Beams (FRIB) it should be possible to extend such studies and to delineate most of the drip line up to mass 100 using the high-power beams available and the highly efficient and selective FRIB fragment separators.

Drip line nuclei often exhibit exotic decay modes. An example is the extremely proton-rich nucleus iron-45 that decays by beta decay or by ejecting two protons from its ground state. Another example of exotic decay modes, proton-rich nuclei exhibiting "superallowed" beta decays, is discussed in "Fundamental Symmetries," later in this chapter. Moving toward the drip lines, the coupling between different nuclear states, via a continuum of unbound states, becomes systematically more important, eventually playing a dominant role in determining structure. Such systems where both bound and unbound states exist and interact are called "open" quantum systems.

Many aspects of nuclei at the limits of the nuclear landscape are generic and are currently explored in other open systems: molecules in strong external fields, quantum dots and wires and other solid-state microdevices, crystals in laser fields, and microwave cavities. Radioactive nuclear beam experimentation will answer questions pertaining to all open quantum systems: What are their properties around the lowest energies, where the reactions become energetically allowed (reaction thresholds)? What is the origin of states in nuclei, which resemble groupings of nucleons into well-defined clusters, especially those of astrophysical importance? What should be the most important steps in developing the theory that will treat nuclear structure and reactions consistently?

*The Heaviest Elements*

What are the heaviest nuclei that can exist? Is there an island of very long-lived nuclei in the N-Z plane? What are the chemical properties of superheavy atoms? These questions present challenges to both experiment and theory. As discussed earlier, the repulsive electrostatic Coulomb force between protons grows so much in those nuclei with large proton number that they would not be bound except for subtle quantum effects. Theory predicts that stability will increase with the addition of neutrons in these systems as one approaches N = 184 (see Figure 2.5), but there is no consensus about the precise location of the projected island of long-lived superheavy elements and their lifetimes (some are predicted to have lifetimes as long as $10^5$-$10^7$ years.

By using actinide targets and rare stable beams, such as calcium-48, elements up to Z = 118 have been produced and observed. The discovery of a nucleus with Z = 117, with a target of berkelium-249, is a case in point as well as an excellent example of international cooperation in nuclear physics (Box 2.1). Not only did this work discover a new element but new information obtained on the half lives of several nuclei in its decay path provided experimental support for the existence of the long-predicted island of stability in superheavy nuclei. Further incremental progress approaching Z = 118 and beyond is possible, but it requires new actinide targets beyond berkelium, and intense beams of rare stable isotopes such as titanium-50. However, there is a range of options for synthesizing heavy elements with exotic beams. By using neutron-rich radioactive targets and beams a highly excited system can be formed, which would decay into the superheavy ground state via evaporation of the excess neutrons. An area of related importance is the further study of the spectroscopy of the heaviest nuclei possible using reaccelerated beams and large acceptance spectrometers, looking at alpha-decay and gamma-ray spectroscopy up to at least Z = 106.

### Neutron-Rich Matter in the Laboratory and the Cosmos

Neutron-rich matter is at the heart of many fascinating questions in nuclear physics and astrophysics: What are the phases and equations of state of nuclear and neutron matter? What are the properties of short-lived neutron-rich nuclei through which the chemical elements around us were created? What is the structure of neutron stars, and what determines their electromagnetic, neutrino, and gravitational-wave radiations? To explain the nature of neutron-rich matter across a range of densities, an interdisciplinary approach is essential in order to integrate laboratory experiments with astrophysical theory, nuclear theory, condensed matter theory, atomic physics, computational science, and electromagnetic and gravitational-wave astronomy. Figure 2.6 summarizes such linkages in this interdisciplinary endeavor.

In heavy neutron-rich nuclei, the excess of neutrons predominantly collects at the nuclear surface creating a skin, a region of weakly bound neutron matter. The presence of a skin can lead to curious collective excitations, for example, "pygmy resonances," characterized by the motion of the partially decoupled neutron skin against the remainder of the nucleus. Such modes could alter neutron capture cross sections important to r-process nucleosynthesis (discussed further in "Nuclear

## Physics of Superheavy Elements

## Chemistry of Superheavy Elements

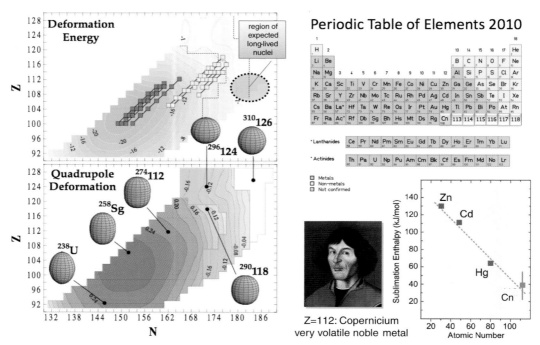

FIGURE 2.5  *Left:* Calculated properties of even-even superheavy nuclei. The upper-left diagram shows the deformation energy (in MeV) defined as a difference between the ground state energy and the energy at the spherical shape. Several Z = 110-113 alpha decay chains found at GSI and RIKEN with fusion reactions using lead or bismuth targets are marked by pink squares and those obtained in hot fusion reactions at the Joint Institute for Nuclear Reactions (JINR) in Dubna are marked by yellow squares. The region of anticipated long-lived superheavy nuclei is schematically marked. *Lower left:* contour map of predicted ground-state quadrupole deformations and nuclear shapes for selected nuclei. Prolate shapes are red-orange; oblate shapes, blue-green; and spherical shapes, light yellow. The symbols $^{274}112$, $^{290}118$, $^{296}124$, and $^{310}126$ refer to unnamed nuclei having the given number of nucleons (superscript) and protons (base). *Right:* Periodic table of elements as of 2010 including the element Z = 112 discovered at GSI and accorded the name copernicium (chemical symbol Cn) in honor of astronomer Nicolaus Copernicus. Its chemistry suggests it is a member of the metallic group 12 (containing zinc, cadmium, and mercury). SOURCES: *(Left)* Adapted by permission from Macmillan Publishers Ltd., S. Ćwiok, P.H. Heenen, and W. Nazarewicz. 2005. *Nature* 433: 705; *(right)* K.L. Jones and W. Nazarewicz, 2010, Designer nuclei—Making atoms that barely exist, *The Physics Teacher* 48: 381. Reprinted with permission from *The Physics Teacher,* Copyright 2010, American Association of Physics Teachers.

Astrophysics," later in this chapter). One of the main science drivers of FRIB is to study a range of nuclei with neutron skins several times thicker than is currently possible. Studies of high-frequency nuclear oscillations (giant resonances) and intermediate-energy nuclear reactions will help pin down the equation of state of nuclear matter.

Another insight is being provided by electron scattering experiments. The Lead Radius Experiment (PREX) at JLAB uses a faint signal arising from parity violation by weak interaction to measure the radius of the neutron distribution in lead-208. This measurement should have broad implications for nuclear structure, astrophysics, and low-energy tests of the Standard Model. Precise data from PREX would provide constraints on the neutron pressure in neutron stars at subnuclear densities. Important insights come from experiments with cold Fermi atoms that can be tuned to probe strongly interacting fluids that are very similar to the low-density neutron matter found in the crusts of neutron stars (see Box 2.2).

## Nature and Origin of Simple Patterns in Complex Nuclei

Rather than tackling the nuclear problem from the femtoscopic perspective of nucleon motions and interactions, one can focus on a complementary view of the atomic nucleus as a mesoscopic system characterized by shapes, oscillations, and rotations and described by symmetries applicable to the nucleus as a whole. In this way, properties and regularities, which might not be explicit in a description in terms of individual nucleons, are highlighted, providing insights that can inform microscopic understanding. Such a perspective focuses on identifying what nuclei do and what that tells us about their structure, while the femtoscopic approach is essential to understanding why they do it.

The mesoscopic approach is motivated by the recognition of, and search for, regularities and simple patterns in nuclei that signal the appearance of many-body symmetries and associated emergent collective behavior. Despite the fact that the number of protons and neutrons in heavy nuclei is rather small, the emergent collectivity they show is similar to other complex systems exhibiting self-organization, such as those studied by condensed matter and atomic physicists, quantum chemists, and materials scientists. While few if any nuclei will exhibit idealized symmetries exactly, such a conceptual framework provides important benchmarks. In this perspective, an important goal is to determine the experimental signatures that spotlight these patterns and the interactions responsible for them. Already, research with exotic nuclei is showing the breakdown of traditional patterns (see discussion of Figure 2.2) and new ways of seeing the emergence of collective phenomena in both light and heavy nuclei.

**Box 2.1**
**U.S. and Russian Scientists Collaborate to**
**Create a New Chemical Element, 117**

A team of U.S. and Russian physicists has created a new element with atomic number Z = 117, filling in a gap in chemistry's periodic table. The new superheavy element, born in a Russian accelerator laboratory at JINR, in Dubna, required coordinated collaborative efforts between four institutions in the United States and two in Russia and more than 2 years to achieve, highlighting what international cooperation can accomplish. The identification of element 117 among the products of the berkelium-249 + calcium-48 reaction occurred in late 2009 and the results were published in April 2010.[1] Production of the berkelium-249 target material, with a short half-life of $T_{1/2} = 320$ days, required an intense neutron irradiation at the High Flux Isotope Reactor (HFIR) of the Oak Ridge National Laboratory (ORNL), chemical separation from other reactor-produced products including californium-252, again at ORNL, followed by target fabrication in Dimitrovgrad, Russia, and six months of accelerator bombardment with an intense calcium-48 beam at Dubna, Russia— a continual intercontinental race against radioactive decay. Analysis of the experimental data was performed independently at Dubna and Lawrence Livermore National Laboratory, providing nearly round-the-clock data analysis by virtue of the 11- to 12-hour time difference between Russia and California. Six atoms of element 117—five of $^{293}117$ and one of $^{294}117$—were observed and 11 new nuclides were discovered in the decay products of those two new Z = 117 isotopes (Figure 2.1.1). The measured half-lives of new superheavy nuclei were observed to increase with larger neutron number. This work represents an experimental verification for the existence of the predicted island of enhanced stability. Scientists and students at Vanderbilt University and the University of Nevada also contributed to this successful experiment.

---

[1] Y.T. Ogannessian, F.S. Abdullin, P.D. Bailey, et al. 2010. *Physical Review Letters* 104: 142502.

FIGURE 2.1.1 Upper end of the chart of nuclides highlighting the 11 new nuclides produced as a result of synthesizing $^{293}117$ and $^{294}117$. The inset shows 22 mg of berkelium-249 in the bottom of a centrifuge cone after chemical separation (green solution). The californium-252 contamination was reduced about $10^8$ times during the purification process at the Radiochemical Engineering Development Center at ORNL. SOURCE: Images courtesy of W. Nazarewicz and K. Rykaczewski, Oak Ridge National Laboratory.

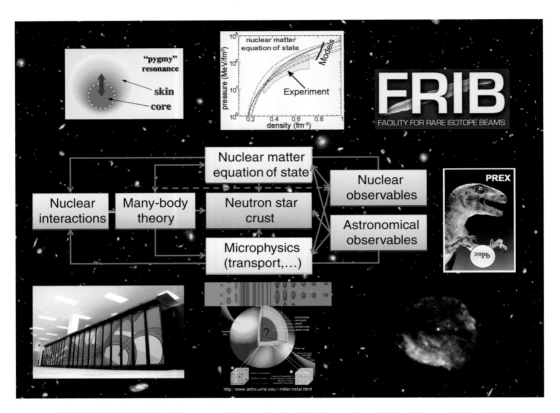

FIGURE 2.6 Multidisciplinary quest for understanding the neutron-rich matter on Earth and in the cosmos. The study of neutron skins and the PREX experiment are discussed in the text. The anticipated discovery of gravitational waves by the Laser Interferometer Gravitational Wave Observatory (LIGO) and the allied European detector Virgo will help understanding large-scale motions of dense neutron-rich matter. Finally, advances in computing hardware and computational techniques will allow theorists to perform calculations of the neutron star crust. SOURCE: Courtesy of W. Nazarewicz, University of Tennessee at Knoxville; inspired by a diagram by Charles Horowitz, Indiana University.

### Nuclear Masses and Radii

The binding of nucleons in the nucleus contains integral information on the interactions that each nucleon is subjected to in the nuclear environment. Differences in nuclear masses and nuclear radii give information on the binding of individual nucleons, on the onset of structural changes, and on specific interactions. Examples of recent measurements of charge radii in light halo nuclei were discussed above. With exotic beams and devices such as Penning and atomic traps, storage rings, and laser spectroscopy the masses and radii of long sequences of exotic isotopes are becoming available, extending our knowledge of how nuclear structure evolves with nucleon number. Figure 2.7 (left) shows the sensitivity of separation energies to nuclear structure. The inset displays the energy required to remove the last two neutrons from the nucleus. These energies have sharp drops after magic numbers but approximately linear behavior in between. Subtracting an average linear behavior therefore magnifies structural changes as seen in the color-coded contours in the two-dimensional plot in the proton-neutron plane.

*Phase-Transitional Behavior*

Changes in nuclear properties as a function of nucleon number can signal quantum phase transitions between regions characterized by different symmetries. Although the behavior of such transitions is muted in finite nuclear systems, experimental studies have provided evidence for their existence and tested simple theoretical schemes for nuclei at the critical points. Theoretical studies that model nuclear shape variations in the limit of large valence nucleon number have shown how phase transitional character in large systems evolved toward the muted remnant of this behavior seen in finite nuclei and helped to identify empirical signatures of first- and second-order phase transitions that have been used to classify the phase transitions in, for example, the A ~ 150 and A ~ 134 mass regions.

The extra binding gained in the shape transition region near N = 90 is evident in the brown shaded area in Figure 2.7 (left). Representative spectroscopic data showing the ratio $R_{4/2}$ of energies of the lowest $4^+$ and $2^+$ excitations are given for the same A ~ 150 region in Figure 2.7 (right). The phase transition is signaled by the concave-to-convex change of pattern between N = 88 and N = 90 associated with a breakdown of a subshell gap at Z = 64.

*Probing Nuclear Shapes by Rapid Rotation*

Gamma-ray spectroscopy is a basic tool for studying nuclear structure, shapes, and their changes—both from the energies and decay paths of excited nuclear states and by measuring nuclear level lifetimes from Doppler effects. Recently, a great diversity of phenomena has been discovered as increasingly sensitive instrumentation reveals unexpected behavior in our quest to observe higher excitation energies and angular momentum states in nuclei.

Figure 2.8 illustrates this progress for the rare earth nucleus erbium-158. The future of gamma-ray spectroscopy is brighter than ever with the development of the next generation of detector systems comprising a highly segmented shell of germanium detectors covering a complete sphere around a source using the new technology known as "gamma-ray tracking." Such systems will have a sensitivity or resolving power about 100 times better than present-day systems. Since gamma-ray spectroscopy is one of the most powerful experimental approaches to unraveling

## Box 2.2
## Intersections of Dense Nuclear Physics with
## Cold Atoms and Neutron Stars

Nuclear systems—from atomic nuclei to the matter in neutron stars to the matter formed in ultrarelativistic heavy ion collisions—are complex many-particle systems that exhibit a great range of collective behavior such as superfluidity. This facet of nuclear systems, shared with matter studied by condensed matter physicists, atomic physicists, quantum chemists, and materials scientists, has opened up splendid opportunities for productive and valuable cross-fertilization among these fields. Of growing importance is the intersection of nuclear physics and ultracold atomic gases.

Atomic gas clouds allow physicists to control experimental conditions such as particle densities and interaction strengths, a control intrinsically unavailable to nuclear physicists. Such control has inspired nuclear physicists to develop more unified pictures of nuclear matter, beyond the constraints of laboratory nuclear systems, and to see commonalities with atomic systems. The experimental flexibility of cold atom systems makes them ideal to explore exotic phases and quantum dynamics in these strongly paired Fermi systems.

The quark-gluon plasmas in ultrarelativistic heavy ion collisions are the hottest materials one can produce in the laboratory, with temperatures of trillions of degrees. On the other hand, clouds of ultracold trapped atoms are the coldest systems in the universe, reaching temperatures as low as one billionth of a degree above absolute zero.1 Nonetheless, despite this difference in temperatures and energies, the two systems share significant physical connections, enabling cross-fertilization between high-energy nuclear physics and ultracold atomic physics. As discussed later in Chapter 2, "Exploring Quark-Gluon Plasma," both systems, when strongly interacting, have the smallest viscosities (compared with their entropy, or degree of disorder) of any system in the universe. The transition observed in strongly interacting cold fermionic atom clouds from paired superfluid states, analogous to superconducting electrons in a metal, to BEC states of molecules consisting of two fermion atoms, captures certain aspects of the transition from a quark-gluon plasma to ordinary hadronic matter made of neutrons, protons, and mesons.

Superfluid pairing in low-density strongly interacting fermionic atomic systems is very similar to that pairing in low-density neutron matter in neutron stars. Figure 2.2.1 compares the predicted energy of a low-density cloud of cold superfluid neutrons with that of cold atomic fermions as the density increases, and shows how the two systems behave in common. Although the energy scales are vastly different, the attractive interactions between fermions in both systems produce extremely large superfluid pairing gaps, on the order of one-third to one-half the Fermi energy, and in this sense these systems are the highest temperature superfluids known. Experiments in cold atoms (illustrated in the inset of Figure 2.2.1) can measure the energies and superfluid pairing gaps of cold fermions from weak to strong coupling, and provide sensitive tests of theories used to compute the properties of matter in the exterior of neutron stars, large neutron-rich nuclei, and quark matter. These properties are key to understanding the limits of stability and pairing in neutron-rich nuclei and the cooling of neutron stars.

One can also study analogues of nuclear and quark-gluon plasma states with cold atoms: Simple examples include the binding of fermionic atoms in three distinct (hyperfine) states, as in lithium-6, analogous to quarks of three colors of quarks, into three-atom molecules, the analogs of nucleons; or the binding of bosonic atoms with fermionic atoms into molecules. One can also exploit similarities of the tensor interaction between nucleons to the magnetic interaction between atoms with large magnetic dipole moments, e.g., dysprosium, to make analogs of the pion condensed states proposed in dense neutron star matter. Strongly interacting ultracold atomic plasmas also present unusual opportunities to study the dynamics of strongly interacting quark-gluon plasmas. Further examples include the formation and interaction of vortices and possible exotic superfluid phases of matter. Future experiments with optical traps will allow one to study the properties of the inhomogeneous matter that exists in the crust of neutron stars. And, strongly interacting clouds of atoms with differing densities of up and down spins, as can be engineered in optical traps, share some common features with strongly interacting quark matter with differing densities of up, down, and strange quarks. In both contexts, superfluid pairing gaps that are modulated in space in a periodic pattern may develop, yielding a superfluid and crystalline phase of matter, hints of which may have been seen in very recent cold atom experiments.

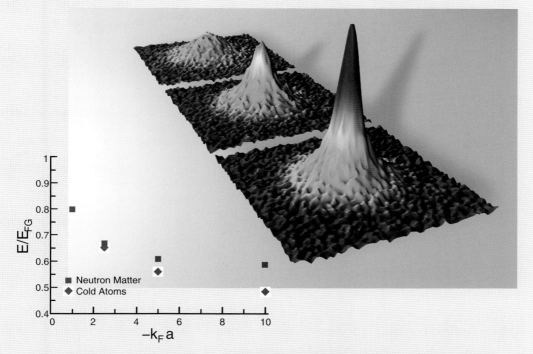

FIGURE 2.2.1 *Upper right:* Images of a superfluid condensate of fermion pairs in the laboratory. The images from upper left to lower right correspond to increasing strength of the pairing obtained by varying the magnetic field. The density tracks the equation of state of strongly paired fermions from the Bose-Einstein condensate (BEC) toward the very strong interaction (or "unitary") limit. *Lower left:* Comparison between the energies of cold atoms and neutron matter at very low densities. The energies are given relative to those of a noninteracting Fermi gas, $E_{FG}$, and are plotted as a function of the product of Fermi momentum, $k_F$, and scattering length a, representing the interaction strength. SOURCES: *(Upper right)* Copyright Markus Greiner, Harvard University, and Deborah Jin, NIST/JILA; *(Lower left)* A. Gezerlis and J. Carlson, *Physical Review C* 77: 032801, 2008, Figure 1. Copyright 2008 by the American Physics Society.

[1] National Research Council, 2007, *Controlling the Quantum World,* Washington, D.C.: The National Academies Press.

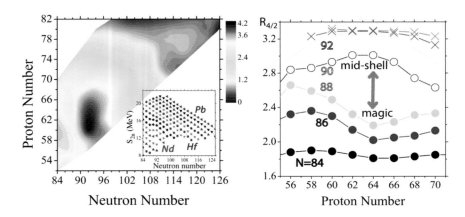

FIGURE 2.7  *Left:* Experimental two-neutron separation energies $S_{2n}$ extracted from measured nuclear masses in the Z = 50-82 and N = 82-126 shells. Removing a smooth reference from the bare values shown in the inset highlights the collective contributions attributed to the valence nucleons. The onset of nonspherical nuclear shapes is clearly seen around N = 90, along with more subtle effects near N = 84 and N = 116. *Right:* Illustration of shape/phase transitional behavior around N = 90. The signature observable $R_{4/2}$ (= E(4$^+$)/E(2$^+$)) varies from <2 for nuclei very near closed shells to ~2 for spherical vibrational nuclei, to ~3.33 for nonspherical nuclei. SOURCE: Figure courtesy of R. Burcu Cakirli, Max Planck Institute for Nuclear Physics, private communication, 2011. Based on data available through 2010.

FIGURE 2.8  *Upper panel:* The evolution of the structure and shape of erbium-158 as this nucleus's rotation speed increases. The excitation energies of various states in erbium-158, with respect to a simple quantum rotor reference, are plotted as a function of angular momentum. A sequence of shape transitions, from weakly deformed prolate, to nearly spherical oblate, to well deformed triaxial is seen. The future challenge will be to reach the region of extremely high nuclear rotations at which erbium-158 cannot withstand the huge centrifugal force and fissions into fragments. The timeline indicates some of the significant milestones in this evolutionary path. *Lower panel:* This timeline, and the rapid development of more sophisticated instrumentation, are further echoed here, where the intensity of a particular gamma-ray transition (normalized to unity for transitions between low angular momentum states) between specific energy levels, as the nucleus de-excites, is plotted as a function of spin. This serves as a guide to the increasing sensitivity as new instruments have become available, starting from single detectors in the 1960s, to small arrays in the 1970s and 1980s, to the current array, called Gammasphere, and to the large gains expected for future generations of gamma-ray tracking arrays such as the Gamma-Ray Energy Tracking Array (GRETINA/GRETA), and the Advanced Gamma Ray Tracking Array (AGATA). SOURCE: Courtesy of Mark A. Riley, Florida State University.

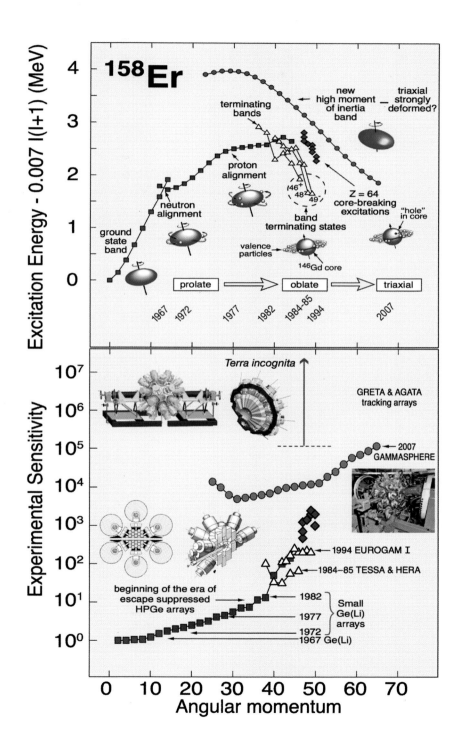

the structure of nuclei, these new, highly sensitive arrays will greatly enhance, for example, the discovery potential of FRIB, which will produce key nuclei—crucial for understanding new structural phenomena of the types discussed on these pages—but often in very small amounts. These prospects are supported by the advances already obtained with existing current-generation instruments.

### New Facets of Nucleonic Pairing

Nucleonic superfluidity plays a large role in nuclear structure. A generic feature of superfluidity is that elementary particles called fermions (such as protons or neutrons) combine to form specially constructed pairs (Cooper pairs) that are bosons and exhibit very different behavior and interactions than their constituent particles. In loosely bound nuclei, pairing may be the decisive factor for stability against particle decay. A striking example is the unbound nature of odd-neutron He nuclei while their even-neutron neighbors are bound. Nucleonic pairing is also important for the structure of neutron star crust. As the number of nucleons can be controlled experimentally, nuclei far from stability offer new opportunities to study pairing. For instance, it has been suggested that, in neutron-rich nuclei, neutron pairs (di-neutrons) are well localized in the skin region. In heavier nuclei with similar neutron and proton numbers, pairing carried by deuteron-like proton-neutron pairs with nonzero angular momentum is expected. Such yet-unobserved correlations are believed to profoundly impact nuclear binding in nuclei with approximately equal numbers of protons and neutrons (N ~ Z nuclei), to influence isospin symmetry and beta decay, and to modify the equation of state of diluted symmetric nuclear matter. Pairing can be probed with a variety of nuclear reactions that add or subtract pairs of nucleons. These reactions can be studied in inverse kinematics (experimental conditions in which the usual roles of target and projectile are interchanged) with a variety of exotic nuclear beams with intensities $>10^3$/s. Because of finite-size effects and different polarization effects in nuclei and nuclear matter, a theoretical challenge will be to relate experiments on nucleonic superfluidity in finite nuclei to pairing fields in neutron stars (see Box 2.2).

A new twist on the pairing story has been provided by studies at the Brookhaven National Laboratory (BNL) and JLAB. These studies precisely probed nuclear interactions on short distance scales, showing that energetic protons are about 20 times more likely to pair up with energetic neutrons than with other protons in the nucleus when nucleons overlap (see Figure 2.9). As discussed earlier, in studies of pair correlations at lower energies, such proton-neutron predominance has not been observed. This can be traced back to variations in the nuclear interaction when changing the relative distance between the two nucleons.

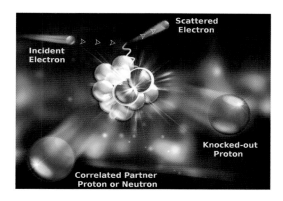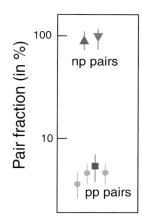

FIGURE 2.9 *Left:* A diagram of a short-range correlation reaction. Knocking out a proton by an energetic electron causes a high-momentum correlated partner nucleon to be emitted from the nucleus, leaving the rest of the system relatively unaffected. *Right:* Depiction of the experimental results from JLAB and BNL that demonstrate the large momentum nucleons in nuclei are primarily coming in proton-neutron pairs. Different symbols and colors mark results of different reactions used. Isolating the signatures of short-range behavior addresses the long-standing question of how close nucleons have to approach before the nucleons' quarks reveal themselves and nucleonic degrees of freedom can no longer be used to describe the system. SOURCES: *(Left)* Courtesy of JLAB; *(right)* adapted from R. Subedi, R. Shneor, P. Monaghan, et al. 2010, *Science* 320: 1476. Reprinted with permission from AAAS.

## Toward a Comprehensive Theory of Nuclei

An understanding of the properties of atomic nuclei is essential for a complete nuclear theory, for an explanation of element formation and properties of stars, and for present and future energy and defense and security applications. Nuclear theorists strive for a comprehensive, unified description of all nuclei, a portrait of the nuclear landscape that incorporates all nuclear properties and forces and can deliver maximum predictive power with well-quantified uncertainties. Such a framework would allow for more accurate predictions of the nuclear processes that cannot be measured in the laboratory, from the creation of new elements in exploding stars to the reactions occurring in cores of nuclear reactors. Developing such a theory requires theoretical and experimental investigations of rare isotopes, new theoretical concepts, and extreme-scale computing, all carried out in partnership with applied mathematicians and computer scientists (see Box 2.3).

There is a well-delineated path toward such a description at the nucleonic level across the nuclear chart that merges three approaches: (1) ab initio, (2) configuration-interaction (CI), and (3) nuclear density functional theory (DFT). Ab initio methods use basic interactions among nucleons to fully solve the nuclear

## Box 2.3
## High-Performance Computing in Nuclear Physics

One of the trends in science today is the increasingly important role played by computational science. Yesterday's terascale computers, capable of a trillion calculations per second, are being replaced by petascale computers, which are a thousand times faster, and scientists are even now working toward exascale computers, which will be a thousand times faster again (at a million trillion calculations per second). All of this computing power will provide an unprecedented opportunity for nuclear science (see Figure 2.3.1). Scientific computing, including modeling and simulation, has become crucial for research problems that are insoluble by traditional theoretical and experimental approaches, too hazardous to study in the laboratory, too time-consuming, or too expensive to solve.

High-performance computing provides answers to questions that neither experiment nor analytic theory can address. As such, it becomes a third leg supporting the field of nuclear physics. Nuclear physicists perform comprehensive simulations of strongly interacting matter in the laboratory and in the cosmos. These calculations are based on the most accurate input, the most reliable theoretical approaches, the most advanced algorithms, and extensive computational resources. Until recently working with petascale resources was hard to imagine, and even at the present time such an ambitious endeavor is beyond what a single researcher or a traditional research group can carry out. To this end, collaborative software environments have been created under the DOE's Scientific Discovery Through Advanced Computing (SciDAC) program, where distributed resources and expertise are combined to address complex questions and solve key problems.[1] In each partnership, mathematicians and computer scientists are collaborating with nuclear physicists to remove barriers to progress in nuclear structure and reactions, QCD, stellar explosions, accelerator science, and computational infrastructure. Computational resources required for these calculations are currently obtained from a combination of dedicated hardware facilities at national laboratories and universities, and from national leadership-class supercomputing facilities.

Although significant advances have been achieved in computer hardware as well as in the algorithms used in today's computations, the forefront computational challenges in nuclear physics require resources that can only be achieved in national supercomputing centers or by dedicated special-purpose machines. Collaborative frameworks such as SciDAC will need to continue in order to prepare for, and to fully utilize, computing resources beyond the petascale when they become available to nuclear physicists. As the nature of the computers will be quite different from that of today's computers, the codes and algorithms will need to evolve accordingly. Given the scale of the computational facilities, it is clear that one should view these numerical efforts like experiments in their style of operation. Currently, the nuclear physics community can efficiently use between 1 and 10 sustained petaflop resources; hence a staged evolution to the exascale seems appropriate.

In summary, the field of nuclear physics is poised to be transformed through the deployment of extreme-scale computing resources. Such resources will provide nuclear physics with unprecedented predictive capabilities that are needed for the systematic exploration of fundamental aspects of nature that are manifested in the structure and interactions of nuclei and

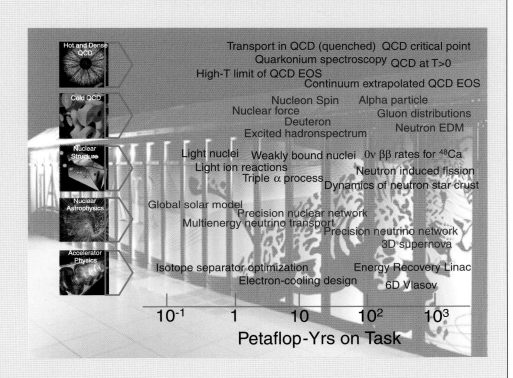

FIGURE 2.3.1 Estimates of the computational resources required to make breakthrough predictions in key areas of nuclear physics: hot and dense QCD, structure of hadrons, nuclear structure and reactions, nuclear astrophysics, and accelerator physics. SOURCE: Adapted from a figure by Martin Savage, University of Washington.

hadronic matter. Future high-performance computing resources will generate enhancements to nuclear physics program that cannot be imagined today.

---

[1] More information about the SciDAC program can be found at http://www.scidac.gov/physics/physics.html; portions of the discussion in this section are adapted from http://www.scidac.gov/aboutSD.html.

many-body problem. Deriving internucleon interactions from quantum chromodynamics (QCD) is a fundamental problem that bridges hadron physics and nuclear structure. While excellent progress has been made in this domain (see the section "The Strong Force and the Internal Structure of Neutrons and Protons"), the lattice calculations have not yet been done with pions as light as those in nature. Meanwhile, QCD-inspired interactions derived within the framework of effective field theory and precise phenomenological forces carefully adjusted to scattering data are commonly used in nuclear structure and reaction calculations. Ab initio techniques have been extended to mass A = 14 and also can be applied to medium-mass doubly magic systems. Configuration-interaction methods adopt the notion of a nuclear potential, which the nucleons themselves both create and move in. This approach has promise up through the region of mid-mass nuclei and heavy near-magic systems. The nuclear DFT focuses on nucleon densities and currents instead of on the particles themselves and is applicable throughout the nuclear chart. The road map for this effort involves the extension of ab initio approaches all the way to medium-heavy nuclei, the development of configuration interaction approaches in a variety of model spaces, and the quest for a nuclear density functional for all nuclei up to the heaviest elements (see Figure 2.10). Special, related challenges are the description of the role of the continuum in weakly bound nuclei and the development of microscopic reaction theory that is integrated with improved structure models.

The nuclear many-body problem is of broad intrinsic interest. The phenomena that arise—shell structure, superfluidity, collective motion, phase transitions—and their connections with many-body symmetries, are also fundamental to fields such as atomic physics, condensed matter physics, and quantum chemistry. Although

FIGURE 2.10 *Top:* The superposed colored bands at the top indicate domains of major theoretical approaches to the nuclear problem. By investigating the intersections between these theoretical strategies, one aims at developing a unified description of the nucleus. *Bottom:* Three examples of theoretical calculations involving high-performance computing—proton densities of carbon-12 obtained in the ab initio quantum Monte Carlo method compared with experimental results *(left)*; numerical simulations of neutrino flavor evolution in a hot supernova environment, where the oscillation probability of neutrinos is shown as a function of neutrino energy and the direction of emission from the surface of a collapsing star *(middle)*; two-dimensional total energy of fermium-258 (in MeV) using nuclear DFT explaining the phenomenon of bimodal fission observed in this nucleus, where nuclear shapes are shown as three-dimensional images that correspond to calculated nucleon densities *(right)*. SOURCES: *(Top; bottom left)* Universal Nuclear Energy Density Functional SciDAC (DOE's Scientific Discovery Through Advanced Computing) Collaboration; *(bottom middle; bottom right)* Nuclear Physics Highlights, Department of Energy; available at http://science.energy.gov/~/media/np/pdf/docs/nph_basicversion_std_res.pdf. Last accessed on April 12, 2012.

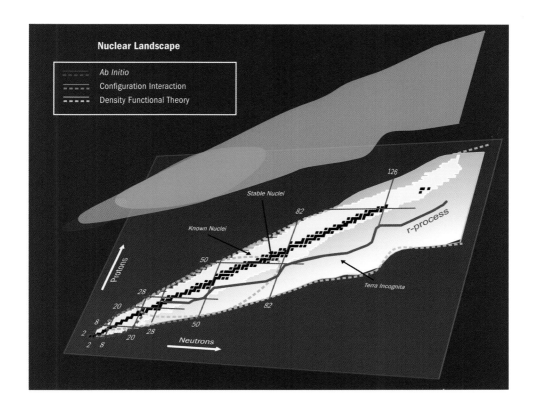

**Nuclear Landscape**

- Ab Initio
- Configuration Interaction
- Density Functional Theory

126

82

Stable Nuclei

r-process

Known Nuclei

50

Protons

28

Terra Incognita

20

8

2

50

8

2   8   20   28

Neutrons

82

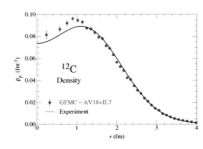

$^{12}C$ Density

GFMC – AV18+IL7

Experiment

$\rho_p$ (fm$^{-3}$)

0.10

0.08

0.06

0.04

0.02

0.00

0   1   2   3   4

r (fm)

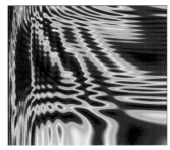

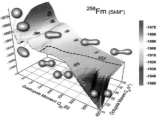

$^{258}Fm$ (SkM*)

the interactions of nuclear physics differ from the electromagnetic interactions that dominate chemistry, materials, and biological molecules, the theoretical methods and many of the computational are shared. Figure 2.10 gives selected examples of many-body calculations.[1]

## NUCLEAR ASTROPHYSICS

The aim of nuclear astrophysics is to understand those nuclear reactions that shape much of the nature of the visible universe. Nuclear fusion is the engine of stars; it produces the energy that stabilizes them against gravitational collapse and makes them shine. Spectacular stellar explosions such as novae, X-ray bursts, and type Ia supernovae are powered by nuclear reactions. While the main energy source of core collapse supernovae and long gamma-ray bursts is gravity, nuclear physics triggers the explosion. Neutron stars are giant nuclei in space, and short gamma-ray bursts are likely created when such gigantic nuclei collide. And last but not least, the planets of the solar system, their moons, asteroids, and life on Earth—all owe their existence to the heavy nuclei produced by nuclear reactions throughout the history of our galaxy and dispersed by stellar winds and explosions.

Among the open questions that will guide nuclear astrophysics in the coming decade are these:

- How did the elements come into existence?
- What makes stars explode as supernovae, novae, or X-ray bursts?
- What is the nature of neutron stars?
- What can neutrinos tell us about stars?

Answering these questions requires understanding intricate structural details of thousands of stable and unstable nuclei, and so draws on much of the work described in the preceding section on nuclear structure. This can be seen in Figure 2.11, which illustrates the principal nuclear processes that shape the visible universe. Each step of each process depends on the nature of that particular nucleus. As an example, a small change of just 10 percent in the energy of a single excited state of one particular nucleus, the famous Hoyle state in carbon-12, would make heavy elements, planets, and life as we know it disappear.

Unraveling the nuclear physics of the cosmos, therefore, requires a broad range of experimental and theoretical approaches. In the last decade, ever more sensitive laboratory measurements of low-energy nuclear reactions enabled precise solar models revealing a deficit of solar neutrinos detected on Earth. Knowledge of this

---

[11]Portions of this paragraph are adapted from Department of Energy, 2007, Computing Atomic Nuclei, *SciDAC Review* 6:42.

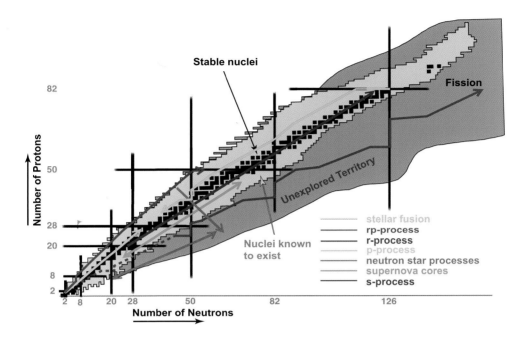

FIGURE 2.11 Schematic outline of the nuclear reactions sequences that generate energy and create new elements in stars and stellar explosions. Stable nuclei are marked as black squares, nuclei that have been observed in the laboratory as light gray squares. The horizontal and vertical lines mark the magic numbers for protons and neutrons, respectively. A very wide range of stable, neutron-deficient, and neutron-rich nuclei are created in nature. Many nuclear processes involve unstable nuclei, often beyond the current experimental limits. SOURCE: Adapted from a figure by Frank Timmes, Arizona State University.

deficit of solar neutrinos combined with the results of advanced neutrino detectors led scientists to the discovery that neutrinos have mass (as discussed in more detail late in this chapter under "Fundamental Symmetries") and confirmed the accuracy of solar models. Laboratory precision measurements also revealed that the nuclear reactions that burn hydrogen in massive stars via the carbon-nitrogen-oxygen (CNO) cycle proceed much more slowly than had been anticipated, changing the predictions for the lifetimes of stars. A few key isotopes in the reaction sequence of the rapid neutron capture process (r-process) responsible for the origin of heavy elements in nature have now been produced by rare isotope facilities. Advanced experimental techniques also enabled measurements of the nuclear properties that characterize their role in the r-process, despite short lifetimes and small production quantities. The same sensitive techniques enabled precision mass and decay measurements of the majority of the extremely neutron-deficient rare isotopes in the rapid proton capture process powering X-ray bursts. The results explain the existence of two classes of X-ray bursts, short and long bursts. In addition, a new rare class of X-ray bursts, so-called superbursts, were discovered and nuclear physics provided the likely explanation of a deep carbon explosion. New multidimensional core collapse supernova models included much more realistic weak interaction physics and nuclear matter properties owing to new results from laboratory experiments and nuclear theory calculations. Contrary to earlier work, some of these supernova models do now explode although many questions about the explosion mechanism remain. In these supernova explosion models, a new type of nuclear process producing heavy elements, the so called neutrino-p process, was found. The discovery of the most massive neutron star to date has eliminated many theoretical predictions about the nature of nuclear matter.

Future nuclear astrophysics efforts are emerging along two frontiers: (1) the study of unstable isotopes that exist in large quantities inside neutron stars and are copiously produced in stellar explosions but difficult to make in laboratories and (2) the determination of extremely slow nuclear reaction rates, which are important for the understanding of stars. Enabled by technical advances, dramatic progress is expected in the coming decade at both frontiers. The FRIB facility in the United States will, together with other rare isotope laboratories around the world, provide unprecedented access in the laboratory to the same unstable isotopes that play crucial roles in cosmic events. And a new generation of high-intensity stable beam accelerators to be located deep underground, as has been proposed for the United States, will enable the measurement of extremely slow stellar nuclear reactions without disturbance from cosmic radiation.[2]

A precision frontier also has emerged in the area of measuring neutron-induced reaction rates using neutron beams. Work is needed at this frontier not only on understanding the origin of those elements produced by neutron capture reactions, but also on applications of nuclear science that depend on neutron capture processes. These applications include the design of novel nuclear reactors and stockpile stewardship, as discussed in Chapter 3.

Nuclear theory is of special importance for nuclear astrophysics for many reasons:

- The extreme densities and temperatures encountered inside stars alter the properties of nuclei compared to what is measured in terrestrial laboratories.

---

[2] Such a facility would also facilitate research in fundamental symmetries, as discussed later in this chapter under "Fundamental Symmetries," as well as in NRC, 2012, *An Assessment of the Science Proposed for the Deep Underground Science and Engineering Laboratory (DUSEL)*, Washington, D.C.: The National Academies Press.

Nuclear theory is needed to calculate the necessary corrections, such as thermal excitations and electron screening.

* In some astrophysical environments such as the r-process or the interiors of neutron stars, extremely rare isotopes exist that cannot be produced in sufficient quantities to fully characterize their properties even with the most powerful rare isotope facilities on the horizon. Experimental data on rare isotopes are needed to advance nuclear theory models, which can then be used to predict the remaining data still out of reach of experiments.

* Many astrophysical reaction rates cannot be measured directly because the rates are too small and the beams too weak. Indirect techniques, where a faster surrogate reaction is used to constrain the slow astrophysical reaction, require reliable reaction theory. In addition, nuclear theory is needed to calculate reaction rates where no experimental information exists.

* Dense nuclear matter can be produced in the laboratory for short times, but can only be observed indirectly from the resulting particle emission. A significant theory effort is necessary to interpret laboratory reaction measurements, and experimental constraints must be used to advance the reliability of the nuclear matter equation of state needed in many astrophysical scenarios.

Progress in nuclear astrophysics must also go hand in hand with progress in astrophysics and observational astronomy. Astronomical observations of the manifestations of nuclear processes in the cosmos provide the link between laboratory and nature. The last decade has seen extraordinary progress in astronomy, with high-precision observations of the composition of very old stars at the largest telescopes on Earth and in space and with surveys scanning hundreds of thousands of candidate stars to find the targets. A new generation of X-ray space telescopes has opened up a novel era in the understanding of phenomena related to neutron stars. Gamma-ray observatories detected the decays of rare isotopes in space, ejected by stellar explosions. Neutrino telescopes provided neutrino images of the sun and had earlier registered neutrinos from a nearby supernova. In the coming decade this progress is bound to continue. Any ongoing large-scale surveys to search for old stars will only pan out in the coming decade, and a new generation of larger ground-based telescopes will enable detailed spectroscopy on many of the resulting targets. Existing X-ray observatories will be complemented with new facilities that push observations toward harder X-rays and possibly gamma-rays and will provide new data on neutron stars and stellar explosions. New-generation gravitational wave detectors are expected to detect signals from supernovae and neutron stars for the first time. Neutrino observatories are ready, and with a little bit of luck they might observe a galactic supernova, an achievement that would revolutionize our understanding of such an event. And a new thrust in astronomy toward wide-field and high-repetition surveys is expected to shed new light on supernovae and to lead to the discovery of new, possibly nuclear-powered, transient astrophysical phenomena.

Astronomy, astrophysical modeling, and nuclear physics need to work together to achieve progress in nuclear astrophysics. Communication across field boundaries, coordination of interdisciplinary research, and exchange of data are essential for these fields to jointly address the open questions. The Joint Institute for Nuclear Astrophysics, funded by the Physics Frontiers Center Initiative of the National Science Foundation (NSF), has been critical in forming and maintaining a unique worldwide platform to foster such interdisciplinary collaboration between the different nuclear astrophysics communities.

Finally, it will be important to strengthen efforts to coordinate research across field boundaries, to form broad interdisciplinary research networks that integrate the wide range of required expertise, and to facilitate the exchange of data and information between astrophysics and nuclear physics, and between experiment, observations, and theory. Such interdisciplinary research networks are also needed to attract and educate the next generation of nuclear astrophysicists, who, with emerging new facilities in nuclear physics, astrophysics, and high-performance computing, are likely to make transformational advances in our understanding of the cosmos.

## Origin of the Elements

The complex composition of our world—some 288 stable or long-lived isotopes of 83 elements—is the result of an extended chemical evolution process that started with the big bang and was followed by billions of years of nuclear processing in numerous stars and stellar explosions (see Figure 2.12). The steady buildup of heavier elements in stars by the successive fusion of hydrogen, helium, carbon, oxygen, neon, and silicon marks the beginning of a new round in the ongoing cycle of nucleosynthesis. The freshly synthesized elements are ejected by stellar winds or supernova explosions and then mixed with interstellar gas and dust from which a new generation of stars is born to repeat the cycle.

Nuclear physics provides the underlying blueprint for this chemical evolution by determining the composition of new elements generated in each astrophysical event. Observations of rare iron-poor, hence old, stars, reveal the composition of the early, chemically primitive galaxy and provide a "fossil record" of chemical evolution. By deciphering the structure of the nuclei involved and by advancing observations, we can trace our chemical history back, step by step, perhaps all the way to the very first supernovae that illuminated the universe. This "nuclear archeology" will advance our understanding of the early universe, of the formation of our galaxy, and also of the future of the universe.

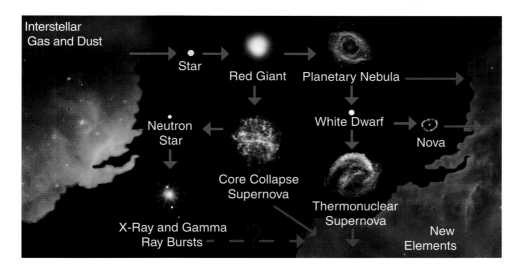

FIGURE 2.12 The ongoing cycle of the creation of the elements in the cosmos. Stars form out of interstellar gas and dust and evolve only to eject freshly synthesized elements into space at the end of their lives. The ejected elements enrich the interstellar medium to begin the cycle anew in a continuous process of chemical enrichment and compositional evolution. SOURCES: *(Background image)* NASA, ESA, and the Hubble Heritage Team (AURA/STScI); *(red giant)* A. Dupree (Harvard-Smithsonian Center for Astrophysics), R. Gilliland (STScI), Hubble Space Telescope (HST), NASA; *(x-ray)* NASA, Swift, and S. Immler (NASA Goddard Space Flight Center); *(planetary nebula)* NASA, Jet Propulsion Laboratory (JPL)-California Institute of Technology (Caltech), Kate Su (Steward Observatory, University of Arizona) et al.; *(thermonuclear supernova)* NASA/Chandra X-ray Center (CXC)/North Carolina State University/S. Reynolds et al.; *(nova)* NASA, ESA, HST, F. Paresce, R. Jedrzejewski (STScI).

*The Eve of Chemical Evolution: How Did the First Stars Burn?*

How were the first heavy elements created by the potentially extremely massive stars formed after the big bang? The pattern of the elements ejected in their deaths might still be observable today in the most iron-poor stars of the galaxy, survivors of an early second generation of stars. Candidate stars with iron content a few 100,000 times lower than that of the sun have been found (see Figure 2.13). Comparing the signatures of these elements with predictions from theoretical models of first stars requires a quantitative knowledge of the nuclear reaction sequences generating these elements. This opens up an observational window into the properties of first stars that is complementary to the planned, very difficult direct observations with future infrared telescopes. The reward might be not only a deeper understanding of the beginnings of chemical evolution in our galaxy but also clues about the nature of the early universe and the formation of structure in the cosmos.

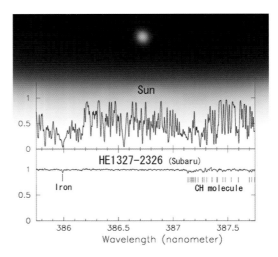

FIGURE 2.13 The most iron-deficient star identified so far has about 400,000 times less iron than the sun. It is believed to be extremely old, having formed shortly after the big bang. The absorption spectrum, here compared to that of the sun, reveals the composition of the early universe at the time the star formed and might contain clues about the elements created by the very first supernovae. SOURCE: *(top)* copyright © Magnum Telescope; *(bottom)* copyright © Subaru Telescope, National Astronomical Observatory of Japan. All rights reserved.

### *Stars: What Elements Are Formed from the Cauldrons of the Cosmos?*

Stars are the nuclear furnaces that forge many of the chemical elements in nature. The composition of the material that stars eject into space depends sensitively on the rate at which the various nuclear fusion reactions occur in their interior. While the reaction sequences have been identified, many reaction rates are still not known accurately, limiting predictions of element formation and stellar evolution. A prominent example is the rate of capture of helium on carbon. With a few exceptions, which mark major milestones in nuclear astrophysics, a direct experimental determination of the low-energy stellar fusion rates has not yet been possible. Some of these pioneering measurements have been enabled by experiments in the low background environments of laboratories deep underground. Models of stars therefore employ uncertain theoretical nuclear reaction rates mostly derived by extrapolating experimental data obtained at higher energies or indirectly.

Addressing this problem will remain a formidable challenge in the coming decade. Advances in experimental techniques such as high-intensity stable beam accelerators in underground laboratories, intense rare isotope beams, and advanced detection and target systems will be needed (see Figure 2.14). On the theoretical

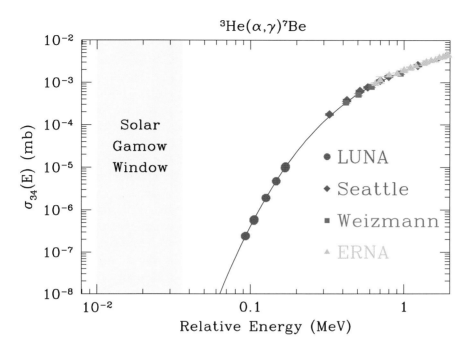

FIGURE 2.14 The fusion probability of helium-3 (3He) and helium-4 (4He), an important reaction in stars affecting neutrino production in the sun, measured directly at various laboratories. The challenge is to measure the extremely small fusion rates at the low relative energies that the particles have inside stars. The reduced background in underground accelerator laboratories (LUNA data shown above in green) compared to aboveground laboratories (all other data) enables the measurement of fusion rates that are smaller by a factor of approximately 1,000. This reduces the error when extrapolating the fusion rate to the still lower stellar energies. SOURCE: Courtesy of Richard Cyburt, Michigan State University.

side, ab initio calculations of nuclear reactions and models that account for cluster structures in nuclei are particularly promising guides for predicting reaction rates at the energies nuclei have in stars. Theory also needs to address the impact of electrons, which always accompany nuclei and modify reaction rates differently in a laboratory target and in stellar plasma.

*The Alchemist's Dream: How Are Gold, Platinum, and Uranium Created in Nature?*

A large gap in our understanding of the chemical evolution of our galaxy surrounds the origin of the elements heavier than iron, such as gold, platinum, or uranium, which comprise more than half of the elements in the periodic table. A slow neutron capture process (s-process) in red giant stars is thought to produce about half of these elements, ending with the production of lead and bismuth. The other half, including the heaviest elements found on Earth, such as uranium and thorium, require an astrophysical environment with an extraordinary density of neutrons. While such an environment has not been identified with certainty, theory predicts that under such conditions, captures of neutrons are very fast, enabling the synthesis of heavy elements beyond bismuth. During the brief duration of this rapid neutron capture process (r-process), exotic short-lived nuclei with extreme excesses of neutrons come into existence as part of the ensuing chain of nuclear reactions. Most of these exotic nuclei have never been made in the laboratory. This will change with the advent of next-generation rare isotope beam facilities like FRIB, which will allow experimental nuclear physicists to produce such nuclei and to determine their properties. The goal is to finally understand how and where nature produces precious metals like gold and platinum and heavy elements like thorium and uranium. Physics questions concerning the neutron-induced processes that constitute the r-process are closely related to neutron-driven applications such as nuclear reactors.

Although the ultimate goal—namely, to identify the astrophysical site of the r-process—has not been reached yet, progress in nuclear physics and astrophysics has been made in the past decade toward unraveling the origin of the r-process elements. Existing radioactive beam facilities have provided experimental data on some of the key nuclei participating in the r-process. Important recent milestones include the half-life measurement of nickel-78 (see Figure 2.15), high-precision ion

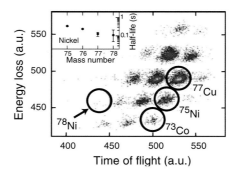

FIGURE 2.15 Very neutron-rich r-process nuclei observed in a rare isotope laboratory. Each dot represents an isotope that has arrived at the experiment, and the dot's location on the map identifies mass number and element. The production and identification of the r-process waiting point nucleus nickel-78 was a challenge, though a sufficient number of isotopes were identified to determine a first measurement of its half-life. Because most r-process isotopes are out of reach of current rare isotope facilities, their study must await a new generation of accelerators such as FRIB. SOURCE: P.T. Hosmer, H. Schatz, A. Aprahamian, et al. 2005. Half-life of the doubly magic r-process nucleus $^{78}$Ni, *Physical Review Letters* 94: 112501 Figure 1. Copyright 2005, American Physics Society.

trap mass measurements of zinc-80, and constraints on the neutron capture rate on tin-132. These data provide guidance for theoretical models, which are used to predict the properties of the many nuclei out of current experimental reach. This has led to recognizing the importance of forbidden beta decay transitions and the direct mechanism in neutron captures and, accordingly, to a more realistic description of nuclear fission.

A variety of astrophysical models have been developed that might provide the conditions necessary for an r-process and eject sufficient amounts of matter into space to account for the observed element abundances. The most promising ones involve core collapse supernovae and the merging of two neutron stars. As a breakthrough, observations of the surface composition of iron-poor stars have opened an unprecedented window into the gradual enrichment of the early galaxy with r-process elements. These stars preserve the composition of the early, chemically less evolved galaxy at the time and location of their formation. The observations tell us that r-process events must have started very early in the evolution of the universe, and that they generate a very robust and characteristic pattern for the abundance of elements throughout the history of the galaxy.

Progress in nuclear physics is needed to connect advances in observations and theoretical astrophysics. In addition to new facilities, the data-driven advances expected in nuclear theory will allow predicting the properties of the nuclei that remain out of reach experimentally and quantifying the errors of such extrapolations. This will reduce the uncertainty in astrophysical models related to nuclear physics to the point where various astrophysical assumptions can be rigorously tested against observations, enabling a data-driven approach to solving the r-process puzzle. New approaches in astrophysics are also needed because none of the existing models achieves the conditions and event frequencies inferred from observations for the r-process. Future large-scale astronomical surveys, followed by high-resolution spectroscopy with the largest telescopes available, need to increase the sample of iron-poor stars formed in r-process-rich environments in the early galaxy to provide statistically relevant information on the frequency of r-process events and the nuclear abundance patterns they produce. Detections of the traces of nearby supernovae in Earth's geological record might also provide clues on the r-process site, and future gamma-ray observatories might be able to detect or at least delimit the radioactive isotopes produced by a supernova r-process.

*Dust Grains from Space: Can They Reveal the Secrets of Stellar Cores?*

The slow neutron capture process is known to occur in red giant stars. But how does matter flow in the deep interiors of stars to generate the necessary free neutrons, and how have these processes changed over the history of chemical evolution? Progress has been achieved in the past decade by analyzing presolar grains—small dust grains that formed in the envelope of a red giant star and travelled through space to be finally incorporated into solar system meteorites. Analyzing the composition of these messengers from space, and comparing them with s-process models that include precise neutron capture rates for stable isotopes measured in an experimental tour de force over many decades, has now led to constraints on the flow of matter in the deep interiors of stars and the dependence of neutron capture rates on galactic age.

In the coming decade experimental data of similar quality need to be obtained for lighter isotopes just slightly heavier than iron, and for so-called branch points. Branch points are unstable nuclei where, depending on nuclear properties, temperature, and neutron density, the reaction sequence splits, producing different isotopes. Once the nuclear properties are experimentally determined, the observed isotopic abundances can be used to infer temperature and neutron density deep inside red giant stars. This is applied nuclear physics par excellence! However, to measure neutron captures on these unstable nuclei, radioactive beam facilities will have to work in concert with neutron beam facilities, where radioactive samples can be quickly irradiated to measure neutron capture rates. Where this is not possible, experimenters and theorists will have to develop new indirect techniques to extract the relevant information from other types of nuclear reactions. The reactions producing neutrons for the s-process are also very uncertain and need to be measured in the coming decades at energies that are closer to the astrophysical conditions than has been possible so far. New radiation detection techniques as well as new high-intensity low-energy accelerators placed in underground facilities to shield experiments from background induced by cosmic rays provide a path forward.

### Blasting Earth with Radioactivity: What Is the Origin of Iron-60?

Neutron captures in stars also produce a long-lived radioactive iron isotope, iron-60, which is ejected in supernova explosions and decays with a half-life of a few million years. Isotopic anomalies found in the solar system indicate that iron-60 was present in the early solar system, and its decay heat might have contributed significantly to planetary melting. Using sensitive nuclear physics techniques, iron-60 has also been discovered in deep sea sediments and on the surface of the moon, possibly indicating an interaction of the solar system with a nearby supernova 2 to 3 million years ago. And the decay radiation of iron-60 has now been detected by gamma-ray telescopes in space. Thus understanding the origin of iron-60 holds the key to learning about conditions inside supernovae, the frequency of supernovae, the possible impacts of a nearby supernova on biological evolution, and the formation of the solar system and planetary systems in general. Developing that understanding requires knowing the efficiency with which nuclear reactions can produce and destroy iron-60 in a given stellar model. Progress has been made on this front by measuring the half-lives and rates of neutron captures on nuclei in the vicinity of iron-60, but the data are still very uncertain. In addition, it has been shown that iron-60 production is sensitive to the rate of various other nuclear reactions governing the evolution of stars, such as the triple alpha process or alpha capture on carbon. Despite decades of effort to measure these rates, the uncertainties surrounding them still prevent a precise prediction of the composition of elements produced in stars.

*Are There Additional New Processes in the Universe Creating Heavy Elements?*

The prevalent view of the origin of the elements heavier than iron and nickel has been that they are made by three distinct processes: the s-process, the p-process, and the r-process. The observations of the composition of old stars show that this traditional picture is not complete as there must be at least one additional nucleosynthesis process producing elements heavier than iron but lighter than most r-process elements in the early galaxy: a so-called light element primary process. The nature of this process remains an open question. At the same time, theory has predicted an unexpected new process producing heavy elements to occur in core-collapse supernovae. During a few seconds of the explosion, hot matter is ejected from the surrounds of the newly born neutron star in the center of the supernova. This matter has a completely unexpected and counterintuitive property: It has more protons than neutrons, caused by interactions with the overwhelming fluence of neutrinos accompanying the explosion. Upon reaching colder temperatures after ejection, nuclei can be formed by combining protons and neutrons. The excess protons can then be captured together with additional neutrons created by proton-antineutrino collisions to produce heavy elements, a process dubbed the νp-process. In the coming decade it will have to be determined if the νp-process and the light element primary process are the same, what their contributions to the chemical evolution of the galaxy are, and what the underlying nuclear physics is. The νp-process involves extremely neutron-deficient rare isotopes, which need to be studied at rare isotope facilities.

## Collapse of a Star

Massive stars end their lives in a violent supernova explosion triggered by the collapse of their cores under their own weight. Core-collapse-induced supernovae can be brighter than billions of stars, and the associated neutrino burst is among the most powerful events in the universe. Such supernovae play a central role in astrophysics. They create and eject most of the elements necessary for life (see Figure 2.16). They are a major energy source driving the evolution of the galaxy by triggering the formation of new stars. And the compact remnants that they leave

FIGURE 2.16 G292.0+1.8, the aftermath of the death of a massive star in a core collapse supernova observed with the Chandra X-ray Observatory. Spectral analysis reveals elements like silicon and sulfur (blue), magnesium (green), and oxygen (yellow and orange) that have been synthesized by nuclear reactions in the progenitor star and during the supernova explosion and have now been ejected into space. The supernova remnant contains a neutron star that has been formed in the supernova explosion and can be observed as a radio pulsar. SOURCE: *(X-ray)* NASA/CXC/Penn State/S. Park et al.; *(Optical)* Caltech's Palomar Observatory. Digitized Sky Survey.

behind—neutron stars and black holes—are the seats of numerous astrophysical phenomena. Yet, it still is not fully understood what makes supernovae explode.

### How Does Stellar Collapse Trigger a Supernova Explosion?

Nuclear physics plays a central role in core collapse supernovae. The collapse of the star's core is powered by gravity but initiated and controlled by the rate at which nuclei are able to capture electrons. With progressing collapse, nuclei in the center are becoming very densely packed, forming a core of nuclear matter that weighs about half of the sun's mass. The repulsive force between the densely packed nuclei halts the collapse and enables an explosion. What happens next is not clear. While the collapsed core stores enough energy to power the supernova, a yet insufficiently known mechanism is needed to transfer that energy to the outer layers of the star and expel them violently.

Solving the supernova puzzle has been a formidable interdisciplinary challenge for many decades, involving (magneto-) hydrodynamics, nuclear physics, particle physics, computer science, and relativity. A complication is that the exploding material moves in turbulent patterns in all directions. This requires multidimensional simulations that push the fastest computers to their limits and beyond.

An important achievement in the last decade was the development of two-dimensional realistic simulations that include the flow and interactions of neutrinos. Such simulations succeeded in predicting explosions of lighter massive stars, albeit not always with the observed features. These explosions were in most cases achieved with the so-called delayed explosion mechanism, where the explosion energy is provided by the strong flux of neutrinos emerging from the compressed, hot stellar core that ultimately becomes a neutron star.

In the coming decade, realistic three-dimensional simulations are anticipated. They will likely tell us whether the simplest neutrino-driven mechanism is sufficient to cause stars to explode and which roles turbulence, rotation, and magnetic fields might play. That is, provided that the underlying nuclear physics is reliably known.

Many of the important rates at which nuclei capture electrons have been significantly improved in recent years by employing modern nuclear structure models. For stable nuclei these rates have been validated successfully using nuclear charge exchange reactions. Such reactions, where an accelerated nucleus interacts with a target in such a way that a proton is exchanged with a neutron, or vice versa, can probe the same nuclear properties that also determine the capture of an electron via the weak interaction. Similar measurements for the many unstable nuclei, which dominate core composition during collapse, will only become possible once rare isotope facilities like FRIB are operational.

The amount of pressure generated by compressing the collapsing core in a supernova to very high density is at the heart of the explosion mechanism. This pressure is determined by the equation of state for nuclear matter at extremely high densities. Current models differ substantially in their predictions, thus producing uncertainties in supernova simulations—for example, concerning the robustness of the explosions obtained. Laboratory measurements and neutron star observations, together with progress in nuclear structure theory, have the potential to significantly reduce these uncertainties in the coming decade.

In the delayed explosion mechanism, the efficiency of energy transfer by neutrinos depends sensitively on the energy spectra of the various neutrino species (flavors), which in turn depend on the weak interactions of neutrinos with the surrounding medium and on neutrino oscillations. In the last decade the understanding of neutrino interactions and their implementation in supernova models has improved dramatically. A recent example is the realization that flavor oscillations induced by interactions of neutrinos with other neutrinos do matter. Neutrino interactions with exotic phases of nuclear matter during the collapse, for example with "nuclear pasta" (described in more detail in the subsection "Neutron Stars"), might also play a role, as does inelastic neutrino scattering on nuclei, which affects neutrino energy spectra.

Observables that directly inform us about the processes in the deep interior of a supernova are difficult to obtain. A nuclear-physics-based diagnostic is radioactive titanium-44, which decays with a half-life of 60 years and whose decay radiation can be detected with gamma-ray observatories (see Figure 2.17), making a detailed

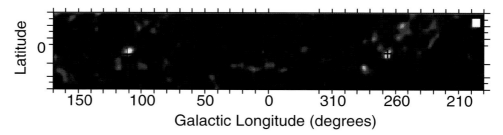

FIGURE 2.17 A map of the galactic plane in 1.157 MeV gamma-rays created by the decay of the radioactive isotope titanium-44. The two locations marked with crosses are statistically significant and coincide with remnants of past supernovae. This enables the detection of past supernovae that are otherwise obscured by dust, and the brightness yields directly the amount of titanium-44 produced, which can be compared with supernova models. The overall low levels of titanium-44 in the galaxy were a surprise, but the nuclear reactions responsible for the synthesis of titanium-44 need to be better studied to draw final conclusions. SOURCE: L.S. The, D.D. Clayton, R. Diehl, D.H. Hartmann, A.F. Iyudin, M.D. Leising, B.S. Meyer, Y. Motizuki, and V. Schönfelder, 2006, *Astronomy and Astrophysics* 450: 1037. Reproduced with permission © European Space Observatory.

understanding of the relevant rare isotope reaction sequences necessary. The future detection of neutrinos or gravitational waves from a nearby supernova might well provide the data needed to answer the question of the explosion mechanism, though with a galactic supernova rate of a few per century one might have to wait for some time. Nevertheless, supernova models must be ready to interpret such observations.

## Thermonuclear Explosions

Thermonuclear explosions of stars can be observed in the cosmos on a daily basis and contribute much to the variability of the night sky as viewed with large telescopes. Even the relatively feeble novae explode with the equivalent of a trillion gigatons of TNT and can often be viewed with the naked eye. The most powerful explosions of this type, thermonuclear supernovae, observationally classified as type Ia, outshine entire galaxies and serve astronomers as distance markers out to the edge of the universe.

Thermonuclear energy is often released in reactions with rare isotopes that do not exist on Earth but are produced under the extreme temperatures and densities arising during the explosion.

### X-Ray Bursts: Can They Be Used as Probes of Accreting Neutron Stars?

X-ray bursts are the most common known thermonuclear explosions of astrophysical origin. A thin layer of hydrogen and helium is accumulated on the surface of a compact neutron star via mass transfer from an orbiting companion star. Typically about once a day, the layer explodes, producing a bright, easily observable burst of X-rays lasting tens of seconds or minutes. In the burning zone of the X-ray burst, hydrogen and helium are completely converted into $10^{16}$ tons of rare isotopes, generating the energy for the explosion. Most of the resulting ashes accumulate on the neutron star surface, where they decay.

The last decade has seen major advances in the understanding of such events. Accelerator facilities that produce rare isotopes have allowed us to measure most of the lifetimes and masses of the very neutron-deficient isotopes produced in the explosions. This has led to important constraints on some of the reactions that generate the burst's energy. Extensive observations with space-based X-ray observatories have discovered rare superbursts, which have been explained theoretically as the reignition of residual carbon in the ashes of the regular X-ray bursts. Yet, many puzzles remain, such as the origin of residual carbon in burst ashes, the nature of multipeaked bursts, occasionally observed very short burst intervals, and light curve anomalies. A particular challenge is to understand bursts well enough to actually extract information about the underlying neutron star, and to predict

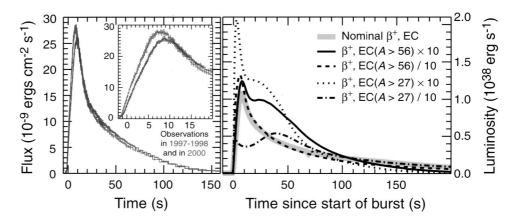

FIGURE 2.18  X-ray bursts observed with the Rossi X-ray Timing Explorer *(left)* from a mass-accreting neutron star and calculated with an X-ray burst model *(right).* The observed burst shapes are found to change over many years of observation, probably due to changes in flow speed of matter onto the neutron star. A quantitative interpretation of the observations is limited by nuclear physics uncertainties, illustrated by the different model predictions of the light curves when assuming different properties of the neutron-deficient rare isotopes responsible for the energy generation. SOURCE: DOE/NSF, Nuclear Science Advisory Committee, 2007, *The Frontiers of Nuclear Science: A Long Range Plan.*

the composition of the ashes that accumulate on the neutron star surface over time and affect many neutron star observations (see the subsection "Neutron Stars"). While a vast amount of observational data is being collected by Earth-orbiting X-ray observatories, the knowledge of many nuclear fusion reactions needed to interpret the observations is still sparse owing to the limited beam intensities of existing rare isotope facilities for the very neutron-deficient isotopes made in X-ray bursts (see Figure 2.18). This is expected to change in the next decade with the advent of a new generation of rare isotope beam facilities where most of the relevant reactions can be measured.

### Novae—Are They Sources of Cosmic Radioactivity?

Novae are an astrophysical phenomenon known since ancient times, when bright new stars suddenly appear in the night sky only to disappear again after a few months. They are now understood to be explosions of a thin hydrogen and helium layer accumulated on a compact white dwarf star via mass transfer from an orbiting companion star. Unlike X-ray bursts, novae emit a lot of visible light. On the other hand, unlike X-ray bursts, novae eject their nuclear ashes into space, possibly explaining the significant amounts of carbon-13 and nitrogen-15 isotopes found today on Earth.

In the last decade, measurements were made of many of the important nuclear rates for novae in a concerted effort at various rare isotope and stable beam facilities. It was observed that some novae produce unexpectedly large amounts of heavier elements such as sulfur. Unfortunately the rates of a few key reactions that produce such heavier elements in novae are still unknown, preventing a quantitative interpretation of these observations. The reaction rates that are particularly difficult to measure involve rare isotopes and will challenge rare isotope experimentation in the coming decade. The situation is similar for the production of radioactive fluorine-18, sodium-22, and aluminum-26 in novae. The gamma-radiation from the decay of these isotopes in nova ejecta might become detectable with next-generation gamma-ray observatories. Alternatively, the presence of these rare isotopes might be revealed in spectral line shifts detected by infrared telescopes.

### What Triggers Thermonuclear Supernovae?

Thermonuclear supernovae are the most powerful thermonuclear explosions in the cosmos. They are believed to consume an entire white dwarf star as fuel, dispersing the ashes into space. Their observed brightness is largely powered by the radioactive decay of nickel-56 into, ultimately, iron-56. This makes thermonuclear supernovae one of the main sources (next to core-collapse supernovae) of iron in the universe. The empirical relationship found between light curve shape and absolute brightness enables astronomers to calibrate thermonuclear supernovae and to use the observed brightness as an indicator of their distance, forming a measuring stick out to the edges of the universe. Indeed, as the 2011 Nobel prize in physics recognized, supernova measurements are at the heart of the new cosmological paradigm of an accelerating universe composed mainly of dark energy.

Yet, what triggers the explosion of the white dwarf star and how the explosion rips through the star, producing the observable distribution and composition of the ejecta, is still unclear. The nuclear fusion reactions that power the explosion are also not understood. The rates of these reactions have to be inferred from data at higher energies where cross sections are sufficiently high for measurements with current techniques. Recent experimental and theoretical progress indicates surprisingly large uncertainties in this approach. On the one hand there are hints of an unexpected reduction of fusion probabilities at very low stellar energies, while on the other hand there is speculation about large enhancements due to unknown resonances.

The challenge for the next decade is to push the sensitivity of nuclear physics experiments to enable reliable estimates of these fusion rates. Advances in our understanding of the rate of electron capture by nuclei and advances in the multidimensional modeling of the explosion itself will allow us to explore theoretically the dependence of supernova features on their stellar environment.

Thermonuclear supernovae convert roughly half of their stellar mass into radioactive nuclei. Detecting the decay radiation from these radioactive nuclei would require a next-generation gamma-ray telescope. Owing to the great penetrating power of gamma-rays, this would be a unique opportunity to probe velocity distributions of matter deep inside the explosion. Coupled with the large amount of observational data expected from large-scale surveys of transient phenomena in the next decade, such a telescope would offer the opportunity to validate the various possible progenitor scenarios, yielding the basic understanding of thermonuclear supernovae needed to quantitatively access systematic errors in the measurements of cosmological distances.

## Neutron Stars

In no other area is the overlap between nuclear physics, astrophysics, and condensed matter physics stronger than in neutron stars (see Boxes 2.2 and 2.4 and Figures 2.6 and 2.19). These are gigantic nuclei, somewhat heavier than the sun, but with a radius of about 10 km and an average density much above that of normal nuclei. Some 100 million neutron stars move around in our galaxy alone. Neutron stars can be studied with telescopes rather than with accelerators and detectors, offering the unique opportunity to understand cold nuclear matter on a macroscopic scale.

The equation of state for cold nuclear matter, the relationship between pressure and density, is key to our understanding of neutron stars. This relationship depends on the properties of very neutron-rich nuclei in the outer crust and on the possible existence of exotic types of matter, such as nuclear pasta—nuclear matter intermediate between regular nuclei and essentially homogeneous neutron-rich nuclear matter—and quarks or condensates of particles such as kaons or pions, which might exist in the center of neutron stars. Hyperons, particles that unlike neutrons and protons include strange quarks, may also exist in neutron stars, although their role is poorly understood owing to insufficient knowledge of their interactions with neutrons and protons.

### What Are the Size, Weight, and Temperature of Neutron Stars?

Measuring the basic properties of neutron stars like mass, radius, and cooling rates is one way to constrain the nature of nuclear matter. Progress in astronomy in the last decade has yielded a range of neutron star masses from timing observations of pulsars. Based on current data, neutron star masses vary, lying between 1.2 and 1.97 times the mass of the sun. The unambiguous detection of a single neutron star with a mass nearly twice that of the sun weeds out various theories of dense nuclear matter. Astronomers continue to search for massive neutron stars, and any discovery of a still heavier neutron star would rule out more theories.

Radius measurements have proven to be more difficult. One approach is to observe the cooling surfaces of neutron stars that were heated by decades of X-ray bursts, but the interpretation of these data remains challenging. New attempts for the simultaneous determination of mass and radius have been carried out using data from soft X-ray burst spectra, which depend on both mass and radius. In the coming decade, long-term observations of changes in pulsations from particular double pulsar systems will offer the opportunity to measure the moment of inertia of a neutron star and will give information on its mass and radius.

Another observable through which neutron stars can be studied is their rate of cooling—neutron stars are born hot and cool over their lifetime. The rate of cooling via the emission of neutrinos depends sensitively on the interior composition. Observations that combine temperature measurements with age estimates and models of neutron star atmospheres can yield information about cooling timescales and would constrain models of the cooling processes. An impressive example is the recent discovery of rapid cooling of the neutron star in the Cas A supernova remnant. Remarkably, this neutron star has cooled significantly in just 10 years. This is perhaps the signature of the neutrons in the neutron star forming a superfluid.

The only way (besides neutrino detection) to peek into the deep interior of a neutron star is through gravitational waves. It is hoped that gravitational waves from the collisions between neutron stars, or between a neutron star and a black hole, will be detected at a rate of at least a few per year in the coming decade with ground-based detectors such as LIGO and VIRGO. The data gained would provide additional constraints on neutron star radii. Future gravitational wave observations of neutron star mergers, giant flares, and continuous wave emission from spinning neutron stars have the potential to directly probe the properties of matter in the interior of the star.

*How Do Rare Isotope Crusts Shape Neutron Star Observations?*

The crust of a neutron star is a relatively thin rigid layer with a thickness of a few 100 m (see Figure 2.19). The outer crust is essentially composed of nuclei in the form of very neutron-rich rare isotopes. The inner crust contains more exotic forms of matter such as superfluid neutrons and possibly nuclear pasta, which is the term used to describe nuclear matter forming into rods and sheets instead of the "drops" typical of nuclei.

The crust and its response to external influences can be observed directly. One example of such an external influence is the occasionally observed "giant flares" that occur on the surface of highly magnetized neutron stars and are energetic enough to directly impact Earth's ionosphere over galactic distances. Oscillations

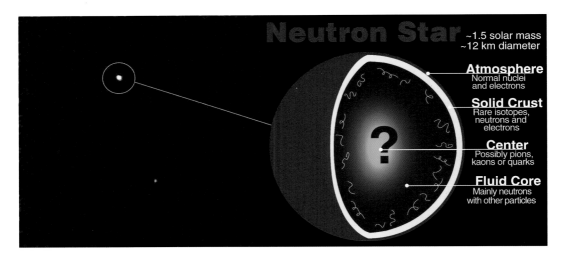

FIGURE 2.19 *(Left)* The neutron star KS1731-260, a giant nucleus in space, observed with the Chandra X-ray Observatory. The observed brightness is heat emitted by the crust, which is a surface layer made of rare isotopes that has been heated by nuclear reactions in an earlier phase of rapid accretion. Accretion stopped in February 2001, and the cooling of the crust has been observed repeatedly since then. *(Right)* Schematic view of a neutron star. The thin crust is mostly made of neutron-rich rare isotopes, while the interior consists chiefly of neutrons with small admixtures of protons, electrons, muons, hyperons, and other particles. At the extreme densities in the core, other forms of nuclear matter, such as a quark gluon plasma, might exist. SOURCES: *(left)* NASA/CXC/Wijnands et al.; *(right)* NASA/GSFC.

observed during these flares are interpreted as starquakes. Much as earthquakes are being used to probe the composition of the crust of the Earth, attempts have been made to use these starquakes to probe neutron star crusts. Similarly, pulsar glitches—sudden changes in the rotation of the neutron star crust, which can be detected with radio telescopes that observe the radio beam emitted by the rapidly spinning neutron star—provide insights into the structure of the crust and evidence for the superfluidity of neutrons. Together these studies have opened up the field of neutron star seismology.

Another example of an external influence is a neutron star that collects a steady flow of gas from an orbiting companion star. This process, together with the various types of thermonuclear bursts that occur in the accumulated gas layer, heats the crust over years or decades. Occasionally this flow of matter gets disrupted. In some cases, modern X-ray observatories have then been able to observe the cooling of the freshly heated crust over many years. Since over time the released heat comes from deeper and deeper layers, the cooling rate contains information about the nature of such layers. Attempts have been made to interpret observed cooling rate changes as indicating that neutrons in the crust are in a superfluid state.

Accumulating matter on the crust of a neutron star from a companion can also induce reactions of rare isotopes that result in density variations, and it can form magnetically confined mountains. Because of the rapid spins of typical neutron stars, both effects are predicted to cause potentially detectable gravitational wave radiation.

The actual crust composition of such mass-accumulating neutron stars depends on the ashes from thermonuclear explosions on the surface, on the rate of electron capture by nuclei, on the properties of the neutron-rich nuclei produced by these captures, and on the rate of a special class of fusion reactions that occur at high density. Fundamental insight into these questions can be gained with the next generation of rare isotope accelerator facilities, which will be able to produce many of the rare isotopes in neutron star crusts.

*Can "Neutron Star Matter" Be Studied in the Laboratory?*

Unfortunately it is not possible to create neutron star matter—cold and extremely neutron-rich nuclear matter—in the laboratory. It turns out, however, that heavy nuclei exhibit a sizeable outer layer that consists predominantly of neutrons—a neutron "skin"—which can be studied in experiments (see "Nuclear Structure" section). The study of hypernuclei (short-lived nuclei where some protons or neutrons are replaced by hyperons) can also provide useful information. These approaches probe neutron star matter at or below the density of matter in the nucleus.

However, deep inside a neutron star much higher densities are reached. Nuclear matter at these higher densities can also be produced in the laboratory, albeit for short times, during the collision of two heavy nuclei. While the dense nuclear matter created by collisions has much higher temperatures than in neutron stars, the key question of how the equation of state of nuclear matter depends on the degree of neutron excess, and how this dependency changes with density, can still be addressed through such collisions. The approach is to identify signatures in the reaction products that probe proton-neutron asymmetry and study them for collisions of nuclei with varying amounts of neutrons. Rare isotope beams will probe these dependencies over a wide range of neutron-proton asymmetry.

## Neutrino Messengers

Neutrinos hardly interact with matter. For this reason they can easily escape from deep interiors of stars like our Sun, from supernovae, and from the interior of our own planet, Earth. The observation of stellar neutrinos constitutes a unique opportunity to look deep into stars and probe extreme astrophysical environments that cannot be simulated in laboratory experiments. However, the very same property that turns neutrinos into messengers from stellar interiors, their extremely small probability of interacting with matter, also makes them extremely hard to detect. Neutrino observatories on Earth therefore require extraordinarily large detectors in underground sites. In the last decades such detectors have been operated with spectacular success, initiating the field of neutrino astronomy. The importance of nuclear physics in addressing fundamental questions about the properties of neutrinos is discussed in detail later in this chapter under "Fundamental Symmetries." At this point we focus on how neutrino observations can be used to shed light on open questions in nuclear astrophysics.

### Can the Sun Be Used as a Calibrated Neutrino Source?

The solution of the solar neutrino problem through neutrino detection and through precision laboratory measurements of the nuclear reaction rates powering the sun is a triumph of nuclear astrophysics (see the section "Fundamental Symmetries"). The goal now is to improve our knowledge of solar hydrogen burning to a level that effectively turns the sun into a calibrated neutrino source. This will require knowledge of the rates of nuclear fusion in the sun to an accuracy of a few percentage points, which will require major advances in the experimental determination of these rates—for example, through accelerators in underground laboratories. Earthbound neutrino detectors with special sensitivities to neutrinos from different solar reactions are in place or in the planning stage and will be capable of performing accurate neutrino spectroscopy. This will further refine our knowledge of the fundamental parameters by which the different neutrino types mix in nature, and at the same time it will probe our understanding of the interior of the sun, its composition, its stability over time, and the processes that transport energy to the surface.

### Can Neutrinos Be Used to Peek Inside a Supernova?

In February 1987, a number of neutrinos that travelled over 10 billion times the distance between the sun and Earth were observed by detectors in the United States and Japan to give the first indication that in the Large Magellanic Cloud a star, Sanduleak-69 202, had exploded as a supernova. This observation was the birth of extrasolar neutrino astronomy and demonstrated unambiguously the theoretical expectation that supernovae observationally classified as type II are indeed triggered by core collapse and that neutrinos are produced in extraordinarily large numbers.

Several advanced neutrino detectors are now in operation and are ready to observe neutrinos from the next nearby supernova with unprecedented detail. Supernova models show that time evolution and energy spectra of supernova neutrinos carry detailed information about the dynamics of the explosion, the explosion mechanism, and the composition of matter in the center of the supernova. A detailed measurement of the supernova neutrino flux as a function of time will be key to understanding the elusive explosion mechanism.

Another opportunity to observe supernova neutrinos, which does not depend on the occurrence of an individual nearby supernova, is the detection of the diffuse background of all neutrinos ever generated by supernovae across the universe that fills the entire cosmos. Detection limits from current neutrino observatories are within an order of magnitude of the theoretically predicted flux, and with further improvements a detection of this background might be possible.

### *Are Neutrino Reactions Responsible for the Existence of Fluorine in Nature?*

The extreme numbers of neutrinos streaming out of the hot core of a developing supernova interact with nuclei in the outer shells of the star about to explode. These interactions happen sufficiently frequently to alter the composition significantly, creating a set of rather rare isotopes of boron, fluorine, lanthanum, and tantalum. The origin of these isotopes is not well understood, but this neutrino-induced nucleosynthesis process provides a natural explanation for their existence. The production rates for these isotopes are quite sensitive to the neutrino energy spectrum and hence temperature. This offers the possibility of using the observed quantities of these isotopes as a unique supernova neutrino thermometer. Doing so would require, besides reliable models of stellar evolution, an accurate knowledge of the interaction of neutrinos with nuclei.

### *Can Neutrinos from Earth's Interior Help in Our Understanding of Earth's Heat Sources?*

More than 4 billion years after its formation in the solar system, Earth's interior is still hot and molten. This heat is largely maintained, it is believed, by the decay of the radioactive elements potassium-40, thorium-232, and uranium-238. Little is known about the distribution and abundance of these elements, but just as with the sun and supernovae, neutrinos emitted upon their decay can be detected with large detectors. Detectors at both KamLAND in Japan and Borexino in Italy have recently observed antineutrinos emanating from Earth's thorium and uranium. An improved understanding of the balance between residual heat of formation and the continuing heat from radioactivity is emerging. More detectors, including one located on the oceanic crust, would help to define Earth's heat sources for geophysical modeling.

*Can the Tiny Interaction Probabilities of Neutrinos Be Measured?*

The rates of neutrino-nucleus interactions inside supernovae and within terrestrial neutrino detectors need to be understood to fully exploit the potential that lies in the detection of neutrinos from astrophysical objects. Theoretical work to determine such interactions needs to be benchmarked with experiments. The challenge is that neutrinos hardly interact with matter. Even with the most intense laboratory neutrino beams and the largest available detectors, direct measurements of neutrino interactions with nuclei are difficult and have been carried out in only a few cases. Charge exchange reactions, where an accelerated ion beam interacts with a target in such a way that a proton is exchanged with a neutron (or vice versa), and reactions where electrons scatter off nuclei probe some of the nuclear physics that determines the rate of neutrino nucleus interactions. These approaches have been successfully used in the past and are expected to be applied to cases of interest in the coming decade at stable and rare isotope beam accelerator facilities. The results will be complemented with predictions from nuclear theory, which can be constrained by the experimental data, to determine astrophysical neutrino-nucleus interaction rates.

## EXPLORING QUARK-GLUON PLASMA

QCD is the theory that describes how quarks and gluons interact. A basic feature of QCD is that although quarks interact strongly when they are separated by distances about the size of a proton, for smaller separations the interaction strength decreases. Physicists realized in the 1970s that this property of QCD implies that the protons and neutrons found in ordinary nuclei can "melt" under extreme conditions, at temperatures above some 2 trillion degrees Celsius. Above this temperature, all matter becomes quark-gluon plasma (QGP) as protons and neutrons merge and release their quark and gluon constituents. For the first few microseconds following the big bang, the entire universe was filled with quark-gluon plasma.

The Relativistic Heavy Ion Collider (RHIC) was built to study the nature of matter at extremely high energy density and to produce states of matter not seen since the universe was microseconds old and then measure their properties. It is now known that by colliding nuclei at very high energies, RHIC creates rapidly expanding droplets of quark-gluon plasma. Experiments at RHIC allow nuclear scientists (in the United States, at 59 universities and 6 national laboratories in 29 states) to answer questions about the microsecond-old universe that cannot be answered by any conceivable astronomical observations made with telescopes and satellites.

Since operations began in 2000, RHIC has provided spectacular evidence that

QGP exists—but it is different than anyone expected. Before 2000, QCD calculations that are known to be reliable at temperatures even higher than those produced at RHIC were used as a qualitative guide to the expected features of the QGP at RHIC. Those calculations suggested that quarks and gluons flew for relatively long distances before bumping into another quark or gluon. If RHIC had created a gaslike plasma of this sort, analogous to familiar electromagnetic plasmas in tokamaks and stars, the produced QGP would have exploded spherically. Instead, the RHIC quark-gluon plasma behaves more like a liquid—in fact, a nearly perfect liquid that flows with very low viscosity. The nonspherical debris patterns from off-center nuclear collisions, and their description as the expansion of a perfect fluid, were the first striking discoveries at RHIC. This liquid QGP is also remarkably effective at slowing quarks as they plow through it, even when the quarks are very heavy and energetic. All observations indicate that the RHIC QGP is not a conventional plasma: There is no evidence at all of any particle-like quark or gluon excitations that travel appreciable distances between interactions. Instead, it is more like a puree-consistency soup than a dilute gas. If it is kicked, its only responses are hydrodynamic waves, like those in water as it reacts to a dropped pebble.

Liquids are characterized by tight coupling between microscopic constituents. As an example, one can consider increasing the coupling in an ordinary liquid such as water. The liquid becomes more "perfect" as the coupling gets stronger, since as the distance that particle-like excitations can travel decreases, a hydrodynamic description becomes ever more accurate and the role of dissipation in damping the flow becomes ever smaller. QGP is a good example of a strongly coupled liquid.

QGP is not the only example of a fluid with no apparent particulate description. The challenge of understanding such liquids appears in several formerly disparate frontier areas of contemporary physics. For example, the interactions among trapped ultracold fermionic atoms, with temperatures around one-millionth of a degree above absolute zero, can be controlled by experimentalists (see Box 2.2). When the interaction is tuned to its maximum strength, so that the atoms travel no appreciable distance between collisions, the atoms behave collectively like a liquid with no particle-like excitations resulting from the underlying atomic degrees of freedom. Indeed, when measuring the shear viscosity of this fluid in appropriate units (by taking the ratio of the shear viscosity to the entropy density), physicists have discovered that this fluid and QGP, almost 20 orders of magnitude hotter, are the two most perfect liquids ever studied in the laboratory (see Figure 2.20). Some of the biggest challenges in condensed matter physics also revolve around understanding phases of matter with no apparent particulate description. Prominent examples include the "strange metal" phase of the cuprate high-temperature superconductors above their superconducting transition temperature, as well as heavy fermion metals containing rare earth elements that are tuned to the vicinity of a zero temperature phase transition, and a lattice of spins in what are known

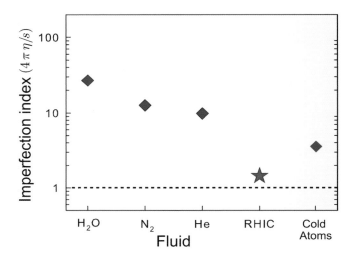

FIGURE 2.20 The ratio of the shear viscosity of a fluid to its entropy density, denoted η/*s*, can be thought of as the "imperfection index" of a fluid, since it measures the degree to which internal friction damps out the flow of the fluid. Quantum mechanics dictates that there are no fluids with zero imperfection index. The two most perfect fluids known are the trillions-of-degrees-hot quark-gluon plasma that filled the microseconds-old universe and that today is produced in heavy ion collisions at RHIC and the LHC, and the millionths-of-degrees-cold fluid made from trapped atoms with the interaction between atoms tuned to the maximum possible. For comparison, the imperfection indices of several more familiar liquids are also shown. Gases have imperfection indices in the thousands. The imperfection index is defined as $4\pi\eta/s$ because it is equal to one in any of the many very strongly coupled plasmas that have a dual description via Einstein's theory of general relativity extended to higher dimensions. Via this duality, $4\pi\eta/s$ =1 is related to long-established and universal properties of black holes. Quark-gluon plasma is the liquid whose imperfection index comes the closest to this value. SOURCE: Adapted from U.S. Department of Energy, 2009, *Nuclear Physics Highlights*, Oak Ridge National Laboratory Creative Media Services. Available at http://science.energy.gov/~/media/np/pdf/docs/nph_basicversion_std_res.pdf.

as "spin liquid phases." In all these cases, the textbook understanding (whether in terms of a dilute gas of quarks and gluons or atoms or Fermi's theory of electrons in a metal) breaks down, failing even at a qualitative level to describe the experimentally observed phenomena. The puzzles raised by experiments in each of these systems are at the core of their respective disciplines. Developing new frameworks for describing such systems represents a fundamental challenge in modern physics that cuts across the boundaries between disciplines.

At very short distances or very high temperatures, the relevant physical laws describing the properties of the QGP are well understood: QCD provides a solid microscopic framework that predicts that QGP does become a dilute gas of particle-like quarks and gluons at very short distance scales and/or very high temperatures.

The challenges and the interest generated by experimental discoveries at RHIC all arise from the fact that in the temperature regime being explored at RHIC, the laws of QCD yield a strongly coupled fluid, rather than a dilute gas of quarks and gluons. Thus, the central challenges in heavy ion physics in the coming decade hinge on detailed investigation of the newly discovered quark-gluon plasma liquid: to quantify its properties, to understand how those properties emerge from the microscopic laws of QCD, and, perhaps most important, to find the right language with which to understand the properties of liquid QGP and with which to gain qualitative insights—insights whose ramifications can then ripple across the many other frontier domains in which strongly coupled liquids with no particulate description present such challenges.

One candidate for a new paradigm to understand strongly coupled fluids goes by the name "gauge/gravity duality." Ideas based on this duality germinated in the late 1990s among string theorists, and in the last decade have bloomed in the hands of both nuclear theorists and string theorists, who together have applied them to the challenges posed by the experiments at RHIC. The basic discovery is that there are many gauge theories ("cousins" of QCD) that feature strongly coupled plasmas in which rigorous calculations can be performed even though conventional methods break down. The key is that in all these examples the quark-gluon plasma turns out to have an equivalent gravitational description in terms of a black hole that lives in four spatial dimensions. One trades the challenges of quantum field theory in three spatial dimensions for classical gravity in one higher dimension, with the extra dimension geometrically encoding the details of how the quarks and gluons interact on different length scales. Via this duality, an extraordinarily small viscosity of the fluid (corresponding to an imperfection index of 1 in Figure 2.20) emerges from a simple and straightforward calculation and it becomes immediately apparent that any fluid that can be described in this way must be nearly perfect. This common feature of so many strongly coupled fluids is related by the duality to a common feature of all black holes—namely, their ability to absorb any object thrown into them and to dissipate any trace of the disturbance. Calculations done via gauge/gravity duality have also yielded qualitative insights into jet quenching (described below) and even predictions for the results of experiments to come. At present, it is not clear whether the qualitative successes of gauge/gravity duality as applied to quark-gluon plasma are just that, or whether they are a sign that QGP itself has a dual gravity description. If the latter were to be the case, quantitative understanding of QGP properties could one day teach us not only about other strongly coupled fluids in condensed matter and atomic physics but also about the nature of the quantum gravitational theory dual to QCD.

Framed by the larger context above, here are some compelling questions raised by the recent discoveries at RHIC that heavy ion collision experiments at

RHIC and at the Large Hadron Collider (LHC) can address in the coming decade:

- The near-perfect liquid QGP discovered at RHIC and now produced also at the LHC must have a particulate description if looked at with a good enough microscope; how, and at what short length scales, can its individual quark and gluon constituents be resolved? And, how does a strongly coupled liquid emerge from constituents that at short length scales are coupled only weakly?
- Experiments at RHIC indicate that the quark-gluon plasma liquid forms and reaches local equilibrium remarkably quickly, in about the time it takes light to travel across one proton. How does this happen? How does the system go from the strong gluon fields hypothesized to occur inside large nuclei to the flowing QGP liquid?
- Does the quark-gluon plasma liquid produced at RHIC and the LHC dissolve even the very small particles formed from heavy quarks and their antiparticles? Does the quark-gluon plasma prevent a heavy quark and antiquark from binding to each other only when they are farther apart than some "screening length"? How close together do they have to be for them to feel the same attraction that they would feel if they were in vacuum?
- How do the energetic particles produced in the earliest stages of a heavy ion collision interact with and deform the fluid? Are very high energy quarks or very heavy "bottom quarks" weakly coupled to the fluid or do they rapidly become part of the soup?

Experiments at RHIC and lattice QCD calculations both indicate that as QGP cools, the reassembly of quarks and gluons into hadrons takes place over a broad temperature range. But, some theoretical calculations indicate that quark-gluon plasma in which there is a greater excess of quarks over antiquarks, as produced in lower energy collisions, should cool through a true phase transition, much like the condensation of water droplets from cooling vapor. If so, there is a sharp phase transition line in the phase diagram of QCD that must end at a critical point. Is there such a critical point in the experimentally accessible domain?

### Discovery of the Near-Perfect Liquid Plasma

Here we go into slightly more depth on various achievements of the last decade, before returning below to the challenge of addressing the questions for the next decade.

*Near-Perfect Liquid*

The distributions of angles and momenta of the end products of a RHIC collision bear witness to the enormous collective motion developed as the tiny drop of fluid produced in the collision expands explosively. In those collisions that are not head-on, the initial droplet is almond-shaped, not spherical. The fluid motion that develops as such a droplet expands is anisotropic—the fluid explodes with greater force about the "equator" of the almond than from its poles, an effect referred to as "elliptic flow" (Figure 2.21). One of the early RHIC discoveries was that a description of these collisions using "ideal hydrodynamics" works surprisingly well, capturing the patterns of how the strikingly large azimuthal asymmetry depends on the impact parameter of the collision and on the identity and momenta of the particles in the final state debris. One of the inputs to ideal hydrodynamics is an equation of state, which is taken from numerical calculations of QCD thermodynamics on a discrete space-time lattice. The other input is the assumption of a perfect liquid

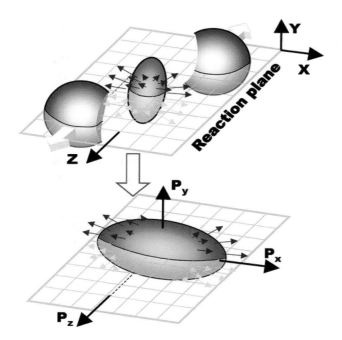

FIGURE 2.21 When the almond-shaped initial geometry in an off-center heavy ion collision at RHIC expands as a nearly perfect fluid, it explodes with greater momentum in the direction in which the initial almond is narrowest. SOURCE: Courtesy of Brookhaven National Laboratory.

with no shear or bulk viscosity (meaning no internal friction that damps out flow). Shear viscosity η enters the hydrodynamic description of relativistic systems in the dimensionless ratio $\eta/s$, where $s$ is the thermal (entropy) density. A convenient figure of merit to quantify the internal friction is $4\pi\eta/s$, called the imperfection index in Figure 2.20. A zero value of the imperfection index is an unachievable idealization like the frictionless inclined plane of high school problems, but in this case its unachievability is a consequence of the laws of quantum mechanics. For typical gases, which are well-described in terms of particles, values of $4\pi\eta/s$ are in the thousands. Terrestrial liquids like water, liquid nitrogen, and liquid helium can have values of $4\pi\eta/s$ as low as 8 to 30 (Figure 2.20). The comparison of RHIC data to recent theoretical calculations done using viscous (nonideal) hydrodynamics demonstrates that the QGP produced at RHIC certainly has $4\pi\eta/s < 5$, and likely has $4\pi\eta/s < 2$. This makes QGP and the strongly coupled fluid made of ultracold fermionic atoms described in Box 2.2 the two most perfect liquids ever studied in the laboratory.

Determining a reliable lower bound on $\eta/s$ and thus quantifying the approach to perfection of the QGP liquid remains a challenge, in part simply because $\eta/s$ is so small. One of the largest, current sources of uncertainty in the value of $\eta/s$ arises from our lack of knowledge of the precise initial density profile of the almond-shaped droplet of fluid formed in the collision. Another uncertainty comes from lack of precise information about how soon after the impact of the two nuclei a hydrodynamic description becomes valid. The success of hydrodynamics indicates very rapid thermalization, but disentangling a precise determination of just how rapid from a precise determination of $\eta/s$ is a challenge. What is needed are additional observables that get at these questions from new angles. Two examples, QCD jets and QGP, are described below.

### Jet Quenching

QCD jets, or directed sprays of particles that emerge from the "hard" large-angle scattering of quarks and gluons within colliding nuclei, are ubiquitous in high-energy collisions of all kinds. Just as differential absorption of X-rays in ordinary matter can be used to explore the density distribution and material composition inside the human body, the absorption of jets in the QGP can be used to obtain direct tomographic information about the properties of the strongly coupled fluid. Jet quenching refers to a suite of experimental observables that together reveal what happens when a very energetic quark or gluon plows through the strongly coupled plasma. It should be noted that these energetic particles are not external probes; they must be produced within the same collision that produces the strongly coupled plasma itself. RHIC was the first facility with energy sufficient to produce these probes in abundance, and even more energetic particles have now been produced in heavy ion collisions at the LHC.

The most pictorial manifestation of jet quenching comes from an analysis in which one looks at the angular distribution of all the energetic particles in an event in which at least one particle with an energy above some threshold has been detected. In proton-proton or deuteron-gold collisions, two back-to-back jets are seen, where each jet is recoiling against the other owing to the conservation of momentum. In gold-gold collisions, however, a single spray of particles around the one used to select the event is seen, but the backward-going jet is missing, as shown in Figure 2.22. Instead, in the backward direction one finds an excess of the much lower energy particles characteristic of the debris from the droplet of QGP itself. The interpretation is that in the selected events one jet emerged relatively unscathed while the recoiling partner quark or gluon plowed into the plasma, dumped its energy into the plasma, and as a result heated the QGP rather than producing a high-energy jet.

Evidence for jet quenching can be seen clearly by measuring the reduction of the number of high-momentum particles observed in heavy-ion collisions. A very energetic quark or gluon loses energy in the QGP predominantly by radiating gluons. How much energy is radiated in gluons is determined by a single material property of the strongly coupled liquid, called the jet quenching parameter, which

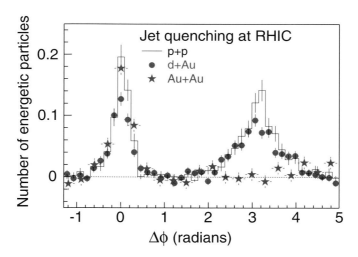

FIGURE 2.22 The extinction of the away-side jet. Collimated patterns of jet particles are observed at $\Delta\Phi = \pi$ opposite a trigger particle at $\Delta\Phi = 0$ in p + p and d + Au collisions but are completely absent in Au + Au collisions. SOURCE: Adapted from DOE/NSF, Nuclear Science Advisory Committee, 2007, *The Frontiers of Nuclear Science: A Long Range Plan.*

is basically a measure of how good the QGP is at slowing the most energetic quarks or gluons shooting through it. Determining the value of this parameter from RHIC data is intrinsically uncertain because the jets studied at RHIC may not be energetic enough to validate the assumptions behind present calculations. It is nevertheless interesting that the jet quenching parameter seems to be larger than it would be in weakly coupled QGP and is comparable to that in the strongly coupled QGP found in QCD-like theories obtained via gauge/gravity duality.

### QGP Shining Brightly

Experimenters at RHIC have recently achieved the long-standing goal of seeing the light (ordinary photons) emitted by the hot, glowing droplets of quark-gluon plasma produced in heavy-ion collisions, as shown in Figure 2.23. What made this a challenge is that there are many more photons produced by the decay of pions (which are formed well after the QGP explodes) than there are photons

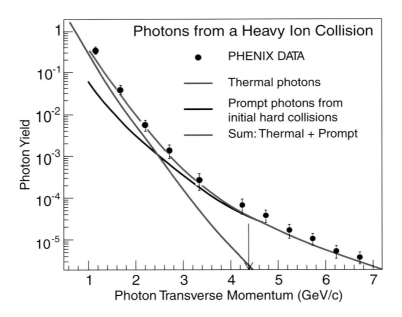

FIGURE 2.23 The measured photon radiation spectrum in Au + Au collisions at RHIC is compared to the spectra expected from photons created by hard collisions in the initial stages of the collision (black curve), from thermal photons radiated by a QGP with an initial temperature (0.33 fm/c after the collision) of 370 MeV (red curve) and from the sum of these processes (blue curve). SOURCE: Courtesy of the PHENIX Collaboration.

from the primordial glowing plasma, and these decay photons must be carefully measured and subtracted. By comparing the spectrum of photons from the plasma to a thermal spectrum—in effect by measuring the color of the luminous glow—experimenters have shown that the time-averaged temperature of the expanding, cooling droplet of quark-gluon plasma is about 30 percent greater than the temperature at which lattice calculations of QCD thermodynamics predict that protons and neutrons melt into QGP. The initial temperature, at the time of thermalization, must be greater still.

*Initial Conditions*

A single head-on collision of gold nuclei at RHIC generates about 5,000 charged particles and 8,000 particles in total. These numbers tell us how much QGP is made in each collision and, with further experimental inputs, also constrain the energy density of the droplets of QGP. Curiously, these strikingly large numbers of particles are lower than had been predicted before RHIC began operations. This observation, together with the suppression of high-transverse-momentum particles at forward angles in deuteron-gold collisions, may indicate that the initial gold nuclei contain fewer low momentum gluons than would be expected from just adding up the contents of independent protons and neutrons. This reduction in the number of gluons, known as "saturation," in turn reduces the number of particles created by gluon-gluon collisions. Saturation results from a characteristic property of gluons in QCD that makes them quite unlike the photons that comprise ordinary light—namely, two gluons can merge into one. The gluon momentum scale below which saturation is thought to arise, denoted $Q_s$, increases with nuclear size and with collision energy, so saturation is expected to be a larger effect at the LHC than at RHIC. Gluons in the incident nuclei with momenta below $Q_s$ are thought to be universal, in the sense that their properties should be the same in nucleons as in nuclei when the collision energy is adjusted so that the two systems have the same $Q_s$. The component of the wave function of nucleons or nuclei that describes gluons in this regime is called "color glass condensate" (CGC). The CGC hypothesis is consistent with measurements of particle yields at forward angles at RHIC and makes predictions for the dependence of the total multiplicity of particles in collisions at higher energy that can be tested in heavy ion collisions at the LHC.

*Novel Particle Production Mechanisms*

Evidence that the droplets of matter formed in heavy ion collisions at RHIC are composed of collectively flowing quarks that are not bound up into protons and neutrons ("deconfined" quarks) comes from detailed measurements of a wide variety of particle species, which reveal surprising patterns in heavy ion collisions

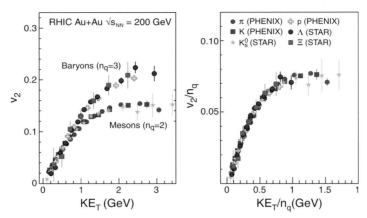

FIGURE 2.24 The strength of the elliptic flow ($v_2$) is plotted as a function of the transverse kinetic energy $KE_T$ for various particle species including mesons (pions and kaons) and baryons (protons, Lambdas, and Xi's) produced in heavy-ion collisions at RHIC (left-hand panel). The right-hand panel shows that the differences between mesons and baryons are eliminated when these quantities are computed on a per-quark basis, strongly suggesting that the flow pattern is determined in the QGP phase. SOURCE: DOE/NSF Nuclear Science Advisory Committee, 2007, *The Frontiers of Nuclear Science: A Long Range Plan.*

that are not seen in more elementary collisions. Species that contain three "valence" quarks (like protons, neutrons, and other baryons) are significantly overabundant at intermediate transverse momenta (2-5 GeV) relative to their abundances in proton-proton collisions. This observation is well described by models in which both baryons and mesons (particles containing one valence quark and one valence antiquark) are produced as a large rapidly expanding droplet of a liquid containing deconfined quarks falls apart (cavitates) into a mist of fine droplets—the baryons and mesons. The baryons and mesons coalesce from quarks drawn from the expanding QGP liquid, a process that is quite different qualitatively and quantitatively from the standard mechanisms by which baryons and mesons form in elementary collisions. Furthermore, the elliptic flow patterns of all baryons of varying masses are the same, and those of all mesons of varying masses are the same, a sign that the elliptic flow was generated during the expansion of the primordial QGP fluid, since then the only thing that matters to the elliptic flow of the observed particles is how many valence quarks each of them took from the primordial fluid. This hypothesis has been beautifully confirmed by scaling the baryon and meson elliptic flow by factors of three and two respectively, as in Figure 2.24, obtaining a universal curve that establishes that the dominant features of the flow pattern are developed at the quark level. It remains an open question how a fluid with no apparent particulate description falls apart in a way that "knows" the number of valence quarks in a baryon or meson.

*Impact of the RHIC Program*

The scientific impact of the RHIC program has been spectacular for a number of reasons. First, RHIC has the ability to produce very large data samples, allowing the possibility of honing in on specific very discriminating observables. Second, the flexibility provided by colliding protons on protons, deuterons on gold, and gold ions on each other opens the way to a natural set of calibration measurements, all made with the same detectors at the same energy. Third, the wide range of available energies allows for a systematic study of observables that identify the properties of QGP and a systematic exploration of the phase diagram of QCD. The combination of flexibility in its design and the dedication of much of its operations to heavy-ion collisions makes RHIC unique among past and present accelerators. RHIC's successes and opportunities can be traced directly to these attributes. Finally, it is also worth noting that the RHIC program and its exciting science discoveries have attracted outstanding young physicists into nuclear physics, further cementing its impact.

## Quantifying QGP Properties and Connecting to the Microscopic Laws of QCD and Its Macroscopic Phase Diagram

In the coming decade nuclear physicists will perform experiments at both RHIC and LHC to address the new scientific questions raised by the RHIC discoveries. The LHC will ultimately achieve heavy-ion collisions with 27 times the collision energy at RHIC, producing QGP that starts out somewhat hotter (perhaps by a factor of two) and that provides probes of this plasma with much higher energy than at RHIC. By 2012, RHIC is expected to reach 20 times its original design luminosity (number of collisions per second). Advances in accelerator physics ("stochastic cooling") will enable RHIC to reach this "RHIC II benchmark" about 4 years earlier than had been envisioned at the time of the 2007 Long Range Plan. Many of the detector upgrades described in 2007 are already in place, and others are anticipated between 2012 and 2014.

Answering the big questions posed above will require measurements at both the upgraded RHIC and the new LHC. First results from lead-lead collisions at the LHC show an elliptic flow pattern nearly identical to that found at RHIC, suggesting little or no diminution of the perfect-liquid phenomena observed at RHIC. If this is confirmed by further measurements and theoretical calculations, the experiments at the LHC will yield the highest energy probes of this liquid. However, the greater flexibility and luminosity at RHIC will give it many advantages in varying the masses and energy of the colliding nuclei to systematically investigate the properties of QGP in various regimes.

*The Search for the Critical Point*

Lattice QCD calculations show that in a matter- antimatter symmetric environment the transition between a gas composed of mesons and baryons to the QGP occurs smoothly as a function of increasing temperature, with many thermodynamic properties changing dramatically but continuously within a narrow temperature range. In contrast, if nuclear matter is squeezed to higher and higher densities without heating it significantly—a feat accomplished in nature in the cores of neutron stars (see Box 2.4)—sharp phase transitions (as in the boiling or freezing of water) may result. A map of the expected QCD phase diagram (Figure 2.25) predicts that the continuous crossover currently being explored in heavy-ion collisions at the highest RHIC energies will become discontinuous if the excess

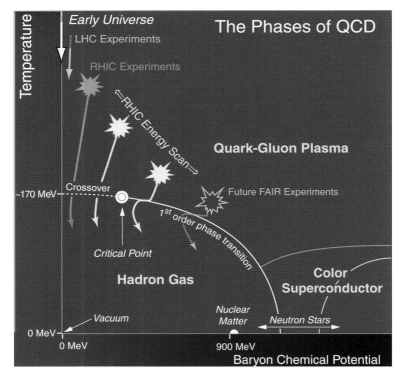

FIGURE 2.25 The phase diagram of QCD is shown as a function of baryon chemical potential (a measure of the matter to antimatter excess) and temperature. A prominent feature in this landscape is the location of the critical point, which indicates the end of the first-order phase transition line in this plane. SOURCE: DOE/NSF, Nuclear Science Advisory Committee, 2007, *The Frontiers of Nuclear Science: A Long Range Plan.*

of matter over antimatter is larger than a certain critical value. This critical point where the transition changes its character is a fundamental landmark on the QCD phase diagram. While new lattice QCD methods have enabled significant progress in the past 5 years toward the goal of locating the QCD critical point, its location remains unknown.

The excess of matter over antimatter in the exploding droplet produced in a heavy-ion collision can be increased by decreasing the collision energy, which reduces the production of matter-antimatter symmetric quark-antiquark pairs and gluons relative to the matter brought in by the colliding nuclei. Decreasing the collision energy also decreases the initial temperature. A series of heavy-ion collision measurements scanning the collision energy can therefore explore the transition region of the QCD phase diagram all the way down to collision energies in which the initial temperature no longer reaches the transition. RHIC completed the first phase of such an energy scan in 2011, taking data at a series of energies to search for a critical point in the phase diagram (if one exists) at a "baryon chemical potential" (a measure of the excess of matter over antimatter) up to about half that in cold nuclear matter. Recent theoretical developments have identified specific event-by-event fluctuation observables most likely to be enhanced in collisions that cool in the vicinity of the critical point. They have also predicted ratios of different fluctuation observables, which, if seen, would fingerprint enhanced fluctuations in the data as being due to the proximity of the critical point. The large range of temperatures and chemical potentials permitted by the flexibility of RHIC, along with important technical advantages in the measurement of fluctuation observables at a collider and the recently upgraded detectors, give RHIC scientists an excellent opportunity to discover a critical point in the QCD phase diagram if, indeed, this landmark falls in the experimentally accessible regime. Later in the decade, the Facility for Antiproton and Ion Research (FAIR) at the GSI Helmholtz Centre for Heavy Ion Research will extend the search to even higher matter over antimatter excess.

*Jet Quenching*

Perhaps the biggest surprise from jet quenching measurements at RHIC is that heavy charm and bottom quarks plowing through the plasma seem to lose energy at a rate comparable to that of light quarks. If the QGP had a particulate description, this would be as surprising as seeing no difference between a bowling ball and a ping-pong ball plowing through a gas of ping-pong balls. Yet data from RHIC on heavy quarks identified indirectly via the isolated electrons produced in their decays show that these "bowling balls" feel the strongly coupled QGP just as much as light quarks do. They not only lose energy, they also seem to be dragged along by the expanding fluid, behaving just like another component of the fluid. At

**Box 2.4**
**Phases of Dense Matter and Neutron Stars**

As the liquid quark-gluon plasma (QGP) that filled the microseconds-old universe expands and cools, it undergoes a phase transition in which it condenses into protons and neutrons, much as steam condenses into droplets of water. This phase transition is merely the beginning: Just as the phase diagram for steam + water + ice (Figure 2.4.1, top) features a rich tapestry of phases and phase transitions, so too does the phase diagram for QCD, illustrated in Figure 2.26 in the main text, in which many phases of matter are expected at high densities and low (relative to trillions of degrees) temperatures. Within just the nuclear matter regime of the phase diagram, where the important axes include the neutron-to-proton ratio as well as the temperature and the density or pressure, varied phases with diverse physical properties emerge (Figure 2.4.1, bottom). Examples include (1) nuclear matter (a liquid of neutrons and protons) with different patterns of superfluid pairing, (2) a dilute gas of deuterons and alpha particles, and (3) rigid crystals made of charged nuclei immersed in a superfluid of neutrons.

Neutron stars are magnificent laboratories within which various phases of dense matter are found. Just below the surface of a neutron star, increasingly neutron-rich nuclei form a rigid crust until a critical density of $4.3 \times 10^{11}$ g/cm$^3$ is reached. Below this depth, the excess neutrons cannot be bound by nuclei and they begin to occupy the space between the nuclei. These unbound neutrons, interacting with one another, become superfluid. Further down, instead of a crystal of nuclei one finds "pasta phases" in which rodlike (noodle-like) and slablike (lasagna-like) structures are embedded in the superfluid. Eventually the pressures are so high that these structures simply merge to form a uniform fluid of neutron-rich matter. This matter (and the neutron fluid above it) are expected to be superfluid only below some density-dependent and therefore depth-dependent critical temperature. It is exciting that observations of the neutron star in the center of the debris from the 320-year-old Cassiopeia-A supernova indicate that it cooled detectably between 2000 and 2010, since cooling that is this rapid could be a sign of the onset of superfluidity, which is expected to happen over decades at different depths.

The properties of matter, as found in neutron stars, at densities significantly greater than that inside large nuclei, have so far eluded fundamental description. What kinds of particles (neutrons and protons? mesons? quarks?) are most important at different densities? And, how does matter make the transition from superfluid neutrons and protons to quark matter at higher densities? Is it sharp or gradual? At higher densities, the usual descriptions of dense matter as a collection of neutrons and protons interacting by static forces must break down: Not only do poorly characterized forces between many neutrons and protons come into play, but a picture only in terms of individual neutrons and protons becomes invalid as the neutrons and protons begin to overlap substantially at higher densities. Indeed, one expects an onset of quark degrees of freedom as these overlaps become important.

FIGURE 2.4.1 Phase diagrams for ice, water, and steam *(top)* and for nuclei and nuclear matter *(bottom)*. SOURCES: *(Top)* Courtesy of Martin Chaplin, London South Bank University. Available online at http://www.lsbu.ac.uk/water/phase.html. Last accessed on April 20, 2012. *(Bottom)* Adapted from and republished with permission of Annual Reviews, from C.J. Pethick and D.G. Ravenhall, 1995, "Matter at Large Neutron Excess and the Physics of Neutron-Star Crusts," *Annual Review of Nuclear and Particle Science* 45: 429, 1995; permission conveyed through Copyright Clearance Center, Inc.

*continued*

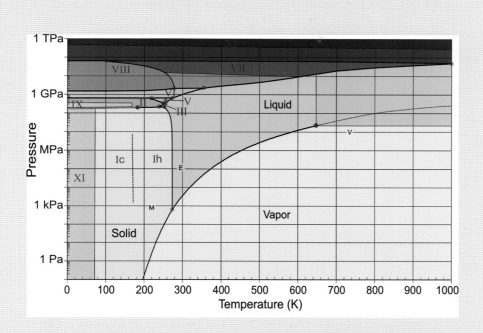

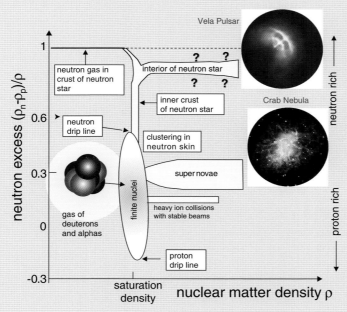

---

**Box 2.4 Continued**

One of the central physical characteristics of nuclear matter is its stiffness—that is, how hard one has to squeeze it in order to increase its density and hence how well it can support matter above it against the overwhelming gravitational forces trying to collapse neutron stars. The stiffness of matter that is a few times denser than nuclei determines the structure of neutron stars, which is governed by a balance between the attraction due to gravity and the stiffness of the matter of which the star is composed. Uncertainties in the properties of matter at densities above those in large nuclei are reflected in uncertainties in the maximum possible mass a neutron star can have, an important factor in distinguishing a possible black hole from a neutron star by measurement of its mass. Astrophysical determinations of neutron star masses and radii strongly constrain the possible stiffness of nuclear matter at high densities. Essentially, the stiffer the matter, the higher the maximum mass that a neutron star can have, but the lower must be the maximum densities found in the centers of neutron stars. Observations of high-mass neutron stars therefore place a lower limit on how stiff matter must be. While it has been known for decades that many neutron stars have masses about 1.4 times that of our sun, the exciting recent discovery of a neutron star with a mass that is reliably determined to be 1.97 ± 0.04 times that of our sun significantly raises the lower bound on the stiffness of nuclear matter and significantly reduces our uncertainty in this fundamental aspect of the phase diagram of QCD. This discovery also tells us that maximum densities in neutron stars do not rise to an order of magnitude above those in laboratory nuclei. Lower maximum densities leave less room for exotic matter in the interior and thus constrain the ways in which neutron stars can cool by emitting neutrinos. The observation of high-mass neutron stars, especially when combined with future neutron star cooling data, presents a deep challenge to our understanding of high density interacting nuclear matter, and at the same time points to the directions that must be taken to make significant advances in solving this outstanding problem.

---

a qualitative level, this is exactly what is expected for a strongly coupled QGP liquid based on calculations of how heavy quarks interact with such a fluid that have been done in QCD-like theories using gauge/gravity duality. The experimental conclusions about heavy quarks are not yet definitive because at present the experiments cannot separate charm quarks from bottom quarks. Upgrades to the detectors in the Pioneering High Energy Nuclear Interaction Experiment (PHENIX) and the Solenoidal Tracker at RHIC (STAR), which are expected to come on line between 2011 and 2014, are designed precisely to address this issue, by separately identifying those particles formed from charm quarks and those formed from bottom quarks.

A second frontier in jet quenching studies is the move from observables based on the measurements of one or two particles to studies of how the angular shape of a jet is modified by the strongly coupled plasma. With the increased luminosity anticipated from 2012 onward, the RHIC experiments will be able to do such measurements for jets with energies more than a hundred times the temperature of the QGP produced in RHIC collisions. At the LHC, such studies have begun using jets with energies that are several hundred times the temperature of the QGP produced in those collisions. It will be very interesting to compare the jet modifications seen in these two regimes and to learn in which regime the effects are larger and in which they can best be measured quantitatively.

At the LHC, the precision of theoretical analyses of jet quenching is expected to improve owing to the higher energy of the jets produced. In addition, new kinds of measurements directly related to LHC's extremely high collision energy will become available. For example, RHIC only recently achieved sufficiently high luminosities to see events in which a single jet is produced back-to-back with one very high energy photon. Such events will be copiously produced at the LHC. The photon flies right through the quark-gluon plasma, allowing the experimenters to measure the initial energy of the jet. Comparison to jets of that same energy in elementary collisions will allow direct measurements of how the strongly coupled plasma modifies the jet.

Studying how the strongly coupled liquid responds to an energetic quark or gluon shooting through it represents a third frontier. There are some indications of a "sonic boom" in the fluid, excited by the supersonic projectile, but the relevant features in the data can also arise from event-by-event fluctuations in the initial shape of the collision zone. Calculations of similar processes in the strongly coupled plasmas of both QCD and the QCD-like theories analyzed via gauge/gravity duality indicate that a quark or gluon can excite a sonic boom, but they also show that more momentum is transferred to the more prosaic wake of moving fluid left behind the energetic probe. The best experimental avenue to resolving these questions is to study collisions between nuclei of different sizes—for example, copper-gold collisions, which can easily be produced at RHIC. The sonic booms, if present, should be similar in these collisions, but the confounding event-by-event fluctuations in the initial shape of the collision zone can be minimized in the head-on collisions of two different-size nuclei.

This many-faceted experimental program, combined with advances in the theoretical modeling needed to interpret the data, will move the field beyond the qualitative conclusions drawn from the present jet-quenching measurements to quantitative statements about essential properties of QGP.

*Dissolving "Quarkonia"*

The hallmark of the QGP is deconfinement, meaning that at high temperatures the quarks and gluons are not confined within baryons and mesons. The very nature of the QGP "screens" the attractive force that normally binds quarks into baryons and mesons. This poses a quantitative question: How close do the quarks have to be in order for their attraction not to be screened? Mesons made from a heavy quark and its heavy antiquark can be used to answer this question because these heavy mesons, referred to generically as "quarkonia," are significantly smaller than protons and typical mesons or baryons. In free space, charm-anticharm mesons (called J/ψ mesons) are roughly half the size of protons, and bottom-antibottom mesons (called Υ mesons) are roughly one quarter the size of protons. It is expected that as the temperature of the QGP increases, one first reaches a regime in which protons and most baryons and mesons have dissolved but the J/ψ and Υ mesons do not fall apart. At higher temperatures, the J/ψ mesons dissolve because the QGP has become hot enough to screen the attraction between a quark and an antiquark separated by the size of a J/ψ, and only at a temperature still higher (by roughly a factor of two) do the smallest mesons known, the Υ mesons, dissolve. Lattice QCD calculations of the screened quark-antiquark potential support this picture, but a definitive experimental confirmation is not yet in hand. Unless obscured by other effects, the sequential deconfinement of these heavy quarkonia should be signaled by a reduction in the number of such mesons detected in the debris of the collision.

This scenario suffers from several complications. For example, given the novel particle production mechanism described above, even if quarkonia dissolve in the QGP liquid, as the QGP cools and reassembles into mesons and baryons, charm and anticharm quarks may find each other and regenerate quarkonia. This confounding complication can be resolved via experimental measurements of the degree to which quarkonia participated in the collective flow of the exploding fluid. This extremely challenging measurement will become possible only with increased luminosity at RHIC. Another concern is that both the J/ψ and the Υ mesons can be excited into larger states, which should dissolve at lower temperatures. Data on both Υ and J/ψ mesons, in heavy ion collisions and in collisions between a proton (or deuteron) and a heavy ion, which serve as a control, can help disentangle this complication. RHIC will have access to (rare)   mesons only after its luminosity upgrade is complete in 2012. It is expected that the LHC (with its higher energy) and RHIC (with its higher luminosity and longer heavy ion runs) will see roughly the same number of J/ψ and Υ mesons per year. The combined analysis of data from RHIC and the LHC should allow comparisons of the production rates for various different quarkonium states in QGP produced with initial temperatures differing by roughly a factor of two. These measurements are demanding, but they have the potential to confirm the pattern of sequential deconfinement and to yield a quantitative measure of the effectiveness with which QGP at varying temperatures can screen the quark-antiquark attraction.

In addition to varying the temperature (RHIC compared to LHC) and the meson size (J/ψ compared to Υ), experimentalists can study quarkonia moving through the strongly coupled quark-gluon plasma at varying speeds. Calculations performed via gauge/gravity duality predict that strongly coupled QGP with a given temperature screens the quark-antiquark attraction more effectively for a quark-antiquark pair that is moving than for one at rest. This indicates that high-velocity quarkonia will dissolve in a lower temperature QGP than quarkonia at rest, making them less numerous. Modeling based on weakly coupled quark-gluon plasma suggests the opposite. High-velocity quarkonia are rare, but the almost completed RHIC luminosity upgrade should bring them within reach, providing us with another measurement that can discriminate between fundamentally different theoretical descriptions of the QGP.

*Uranium-Uranium Collisions*

Since 2011, RHIC has had a new ion source, the Electron Beam Ion Source (EBIS), which makes it possible to accelerate uranium nuclei. These highly deformed nuclei are about 30 percent larger from pole to pole than across their equators. The goal of the initial brief uranium-uranium run anticipated at RHIC in 2012 is to demonstrate that experimentalists can select a subset of all the collisions in which the nuclei collide tip-to-tip, and a different subset in which they collide side-on-side. If this can be done, uranium-uranium collisions in the coming years can play a significant role in answering currently open questions about how perfect the QGP liquid is and about jet quenching. The side-on-side collisions will produce elliptical droplets of QGP with an initial density profile different than in the off-center collisions of spherical nuclei. This opens a path to separating the effects of variations in this density profile from the effects of a small but nonzero $\eta/s$. The tip-to-tip collisions will achieve higher energy densities than current RHIC collisions in a smaller transverse area. Currently, reducing the transverse area by selecting off-center collisions necessarily also reduces the energy density. The comparison between uranium-uranium and gold-gold collisions will allow separate control of the size of the QGP droplet and its energy density, allowing for clean studies of the path-length dependence of jet quenching observables, one of the key discriminants between different theoretical calculations that model the energy loss of energetic quarks moving through the QGP.

*Direct Photon Flow*

How RHIC has seen the glow of the QGP was described above. With the much higher luminosities anticipated in several years, it will become possible to measure the angular distribution of this electromagnetic radiation. Once they are produced, photons do not interact further with the QGP. This means that the angular asymmetry of the light from the glowing QGP will enable experimenters to see earlier stages of the expanding QGP droplet, opening another pathway to a quantitative determination of the figure of merit $\eta/s$.

### Toward a Theoretical Framework for Strongly Coupled Fluids

*Lattice QCD*

Lattice gauge theory is an essential tool for calculating the properties of strongly interacting matter in thermodynamic equilibrium directly from the fundamental theory of QCD. As a result of a series of theoretical breakthroughs and the rapid rise of computing power, lattice QCD has had a significant impact on the understanding of experimental data and static properties of strongly coupled QGP. Lattice calculations like those in Figure 2.26 have determined the equation of state for QCD matter, and hence the temperature of the crossover transition between ordinary matter and quark-gluon plasma, with an accuracy of better than 10 percent. Lattice QCD calculations of the static screening potential between two heavy quarks are an essential phenomenological input to analyses of heavy quarkonia within quark-gluon plasma. Progress has been made using lattice QCD to map out the phase diagram of QCD at nonzero temperature and moderate matter-over-antimatter asymmetry. In the coming decade it should be possible to use lattice techniques to determine the location of the QCD critical point, with important consequences for experimental efforts to detect this important feature of the QCD phase diagram. Lattice calculations of the dynamical properties of strongly coupled QGP are even more challenging. Pioneering calculations of transport coefficients, which describe how the plasma responds to external perturbations, have begun. These calculations address quantities including the shear and bulk viscosities and electrical conductivity of QGP, the diffusion constant for heavy quarks, and the fluctuations of conserved charges and of the density. They are still in their infancy: For example, almost all of them neglect the presence of quarks and antiquarks in the QGP. To have quantitative relevance for RHIC and LHC phenomenology, such calculations must be performed with dynamical light quarks, which will become possible in the coming decade as computing capability approaches the exascale regime.

*Relativistic Dissipative Fluid Dynamics*

In the last several years, nuclear theorists for the first time established a consistent framework for relativistic hydrodynamic calculations that include accounting for the effects of the shear and bulk viscosities. They then developed the codes needed to describe the anisotropic expansion of the exploding droplets of QGP produced in heavy ion collisions. An important microphysical input to these calculations is the equation of state, obtained from the lattice QCD calculations discussed above. These viscous hydrodynamic calculations have become a pillar of quantitative RHIC phenomenology, and they provide the benchmarks for early LHC data. It is these calculations that have made it possible to say that the RHIC

Symmetries," provide powerful tests of the Standard Model in ways that complement the planned program of the Large Hadron Collider (LHC).

## Momentum and Spin Within the Proton

If mass is the first property one thinks of in characterizing a proton, then spin is a close second. Thousands of MRIs performed every day rely on the fact that the magnetic moment of the proton, which is associated with its spin, has an anomalously large value. Yet while the origin of the mass of the proton can at least be postulated, the way in which the proton's spin arises from the dance of the quarks and gluons within it remains unknown. It should be possible to understand the proton's spin through the lens of QCD. Before the predictions of the theory were well understood, it was assumed that since every quark in a proton has the same spin as the proton itself, the proton's spin could be explained by assuming it was made of three quarks, two of whose spins pointed in opposite directions and so cancelled out. If this naïve picture was a good guide, the observed spin of the proton would be due to the spin of its third quark, with the gluons, the sea quarks, and orbital motion all playing unimportant roles. The failure of this picture was first identified at CERN in the late 1980s, where it was discovered that the spin of all the quarks within a proton, when added up, accounted for only a small fraction of the proton's overall spin. Efforts at laboratories around the world, including Brookhaven National Laboratory BNL, CERN, Deutsches Elektronen-Synchrotron (DESY), JLAB, and the Stanford Linear Accelerator Center (SLAC), using a number of complementary techniques, have established that quark spins account for only about 30 percent of the proton spin, and very little if any comes from the spin of the gluons. By elimination, this means that orbital motion must be a key player after all. Figure 2.30 gives an accounting of the proton's spin; identifying the missing pieces has been the subject of substantial effort in the last decade.

Once again, major technical developments in polarized beams and targets, in this case polarized colliding proton beams at RHIC, are enabling the key experiments. The polarized proton beams make it possible to constrain the spin contribution of the gluons to the proton, virtually unknown until now. Figure 2.31 shows results indicating that the gluon spin contributes less than 10 percent of that of the proton over the region in which it has been measured. In the next decade, this limit should be improved dramatically at RHIC as the colliding beam energy ramps up to 500 GeV. The RHIC-spin program has also demonstrated that the charged weak current can provide stringent constraints on the light quark contributions to the total spin. As more data are added to the mix, they will make it possible to determine what fraction of the 30 percent of the proton spin that comes from quarks comes from the sea of quark/antiquark pairs and what fraction comes from the three extra quarks responsible for the charge of the proton. These results are

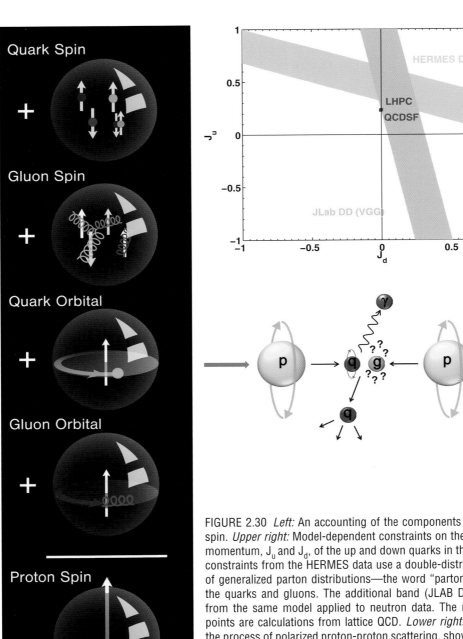

FIGURE 2.30 *Left:* An accounting of the components of the proton spin. *Upper right:* Model-dependent constraints on the total angular momentum, $J_u$ and $J_d$, of the up and down quarks in the proton. The constraints from the HERMES data use a double-distribution model of generalized parton distributions—the word "parton" referring to the quarks and gluons. The additional band (JLAB DD) is derived from the same model applied to neutron data. The red and black points are calculations from lattice QCD. *Lower right:* A cartoon of the process of polarized proton-proton scattering, showing the interaction of a polarized quark with a gluon. SOURCES: *(Left)* Courtesy of J.G. Rubin, Argonne National Laboratory; *(upper right)* Courtesy of K. Orginos, College of William and Mary, the HERMES Collaboration, and JLAB.

forcing scientists to reconsider the contribution of orbital motion to the spin of the proton and should help create a more complete picture of how the dynamics of quarks and gluons cooperate to produce the protons and neutrons that form the visible matter of the universe.

At DESY, the HERMES experiment provided direct evidence that the quarks bound together in the proton have significant orbital angular momentum, which might account for some of what's missing. New experiments have been devised to measure the orbital angular momentum carried by the quarks as well as the gluons. Since orbital motion comes about from a sideways motion relative to the direction of a force, processes that can identify the transverse components of the momentum of quarks and gluons are a key focus of planned experiments.

The HERMES experiment opened a new window into the structure of the nucleon. After the energy upgrade to 12 GeV at JLAB is completed, experiments at the lab will systematically probe the correlated space and momentum distributions of the quarks, exposing the pattern of orbital motion. These new experimental results incorporated in the new conceptual framework being developed by QCD theorists, the so-called "generalized parton distributions," are the best known method to determine the total angular momentum of the quarks in the proton due to their orbital motion.

Another challenge is to determine the contribution of each flavor of quark. Unraveling the proton's flavor structure came to the forefront when, in the 1990s, an unexpectedly large flavor asymmetry in the light quark sea was discovered. Prior to that it had been assumed, based on the simplest estimates from QCD, that the light quark sea had an identical distribution of up and down quarks. A new experiment at Fermilab makes use of the 120 GeV proton beam from its main injector, which turns out to be a nearly ideal energy to study this light quark asymmetry. These measurements will play a role in interpreting data from the LHC at CERN, where experiments will operate in a regime in which detailed knowledge of the distributions of the most energetic quarks in the nucleon is essential.

As discussed in the section "Exploring Quark-Gluon Plasma," the main objective of experiments at the LHC is to test our understanding of the properties of matter at the highest energies. In these experiments, collisions of energetic particles result in showers of new particles whose properties are measured with sophisticated detectors. The recorded results of the collision contain both the signature of the interactions that occurred at the highest energies and the imprint of the strong interaction effects (QCD) that occur as the particles slow down moving away from the center of the collision. Uncovering the high-energy effects requires both ingenious experimental techniques and full control of the theoretical predictions of the Standard Model. Apart from QCD, the other interactions described by the Standard Model are weak, and their consequences are well enough understood to help constrain QCD. The strong interactions, on the other hand, require special

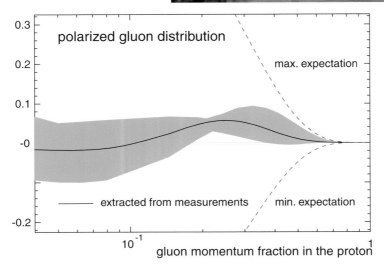

treatment to obtain the required precision. Theorists have shown that lattice QCD calculations can reliably determine the effects of strong interactions needed for obtaining the parameters of the Standard Model, as well as uncovering new phenomena at the frontier of high-energy physics. This is one of the main motivations behind the support of the lattice QCD effort in this country and worldwide. If the experimental findings at the LHC were to show that new, strongly coupled QCD-like theories are needed to extend the Standard Model beyond our current understanding, the lattice methodologies developed for these quantitative QCD calculations could prove to be invaluable at a much higher energy scale.

### In-Medium Effects: Building Nuclei with QCD

How do electron transport properties become modified in a semiconductor, and what does this teach us about the range and nature of the force relevant in complex assemblies? This is a central question in condensed matter physics, essential for an understanding of how the properties of materials exploited for useful electronic devices emerge from the underlying theory of quantum electrodynamics. An analogous question in nuclear physics—How do complex nuclei found in nature or created at accelerators emerge out of QCD?—will be addressed in the next decade (see Figure 1.4). Owing to the nature of the strong interaction, this is indeed a grand challenge, but the first steps toward understanding the construction of nuclei from QCD are being taken through the study of quark-based phenomena in nuclei. For example, one can look for situations where more than one proton or neutron must be involved in elements of the nucleus assembly, or try to understand how quarks behave as they travel through nuclear matter, or how the internal properties of neutrons and protons are affected when they are embedded in nuclei rather than in their free state. Below are some of the important advances in these and other areas within the last decade.

FIGURE 2.31 The PHENIX muon detector upgrade shortly after construction at the University of Illinois at Urbana-Champaign *(top)* and in place at RHIC (NSF-RIKEN collaboration) *(middle)*. Bottom: A global fit to existing data from the PHENIX and STAR detectors at BNL as well as from other experiments around the world. The distribution of gluon angular momentum can be extracted from these data and has been found to be very small. The green band indicates the bounds from the fit; the various dashed lines indicate the ranges allowed prior to the RHIC measurements. SOURCES: *(Top left)* Courtesy of Young Jin Kim, University of Illinois at Urbana-Champaign; *(top right)* Courtesy of Martin Leitgab, University of Illinois at Urbana Champaign, and Larry Bartoszek, Bartoszek Engineering, Aurora, Ill.; *(bottom)* Courtesy of Marco Stratmann, Brookhaven National Laboratory. Adapted from D. de Florian, R. Sassot, M. Stratmann, and W. Vogelsang, 2008, *Physical Review Letters* 101: 072001.

*How Can the Properties of a Proton or Neutron Be Modified in a Nucleus?*

The effects on the properties of a proton or neutron of the medium in which it finds itself are examples of the emergent phenomena associated with an extended system. As such, they fall into a class of inquiry that intrigues scientists in almost all fields of physics and that is responsible for lucrative practical applications such as the electronic devices that emerged from the field of semiconductor physics. In nuclear physics, the EMC effect (named after the European Muon Collaboration that discovered the effect in the 1980s at CERN) is one such phenomenon, believed to arise from modifications of neutrons and protons inside the nuclear medium. Measurements reveal a clear difference between the quark distributions in heavy nuclei compared to the lightest nucleus, deuterium. A central question has been to distinguish conventional nuclear structure effects from processes that are not described by models using only nucleons as the building blocks. One particularly fruitful approach is to focus on light nuclei with between 2 and 12 nucleons, where the nuclear structure is well understood experimentally and well described by existing models. The EMC effect should be most dramatic in the densest nuclei, like helium-4, which has two neutrons and two protons in a tightly bound package. The nucleus beryllium-9 is an example where the average density is only a little more than half that of helium-4 despite its additional five nucleons. It came as a surprise that the observed EMC effect was as large in beryllium-9 as in helium-4 or in carbon-12, which is also tightly bound. A close experimental examination of the nuclear structure of beryllium-9 revealed that it behaves more like two alpha particles (i.e., two dense helium-4 nuclei) and a neutron rather than like nine evenly distributed nucleons, so there is a significant likelihood that the scattering takes place from one of the alpha particles within the beryllium-9 nucleus. This measurement is an important step toward understanding a long-standing puzzle, and new experiments at JLAB and Fermilab, coupled with theoretical developments, are expected to provide the final resolution.

*Can the Strong Interaction Be Weakened at the Femtoscale?*

The transition between the long-distance, low-energy regime, where quarks are strongly interacting and QCD is difficult to apply (the nonperturbative regime), to the short-distance high-energy (perturbative) regime, where approximation techniques similar to those used in quantum electrodynamics (QED) can be applied, is revealed in nuclear reactions. At high energy, the concept of asymptotic freedom applies, in which the coupling strengths decrease and the quark and gluon interactions become weak enough to be described in the approximation of perturbation theory. Apparently at lower energy the quarks acquire mass, and their response to the strong force becomes ever stronger as they accumulate a cloud of virtual quarks and gluons that bind to them with extra heft. This feature is displayed in Figure 2.32. A lattice QCD calculation shows that the effective quark mass is momentum dependent, rising as the momentum becomes lower. Deep inelastic scattering studies on the proton have shown that the transition from the nonperturbative regime to the perturbative regime occurs when the momentum transferred to the quark is between 1 and 2 GeV. As seen from Figure 2.32 this is where the change in the quark mass shows a sharp rise. However, in this low-energy regime, the confining interactions have become so strong that the notion of the quark mass ceases to be well defined. In this regime, the proton is a complicated, many-body system whose mass arises primarily from the interaction energy between its constituents. In Figure 2.32 (right panel), a lattice QCD calculation shows that the proton mass changes very little even when the quarks are assumed to be massless. In the 12 GeV JLAB upgrade, enough energy will be available to allow the probing of this transition between low-energy and high-energy regimes, where quarks and gluon dynamics can be described with analytic methods.

*Can the Transition from Free Quarks to Bound Quarks Be Understood?*

Confinement—the complete absence of free quarks in nature—is a striking and unique property of the strong interaction. A principle effort in nuclear physics is to understand confinement in the context of QCD. In the deep inelastic scattering of an electron off a quark in a nucleus, the struck quark transforms into multiquark bound states, or hadrons, through a process that is not understood and has been only qualitatively described. The central method for studying this hadronization process is to use the nucleus as a laboratory for testing ideas about short-distance behavior. Current thinking posits that a prehadron forms first that is less likely to interact with the nucleus than a bare quark. Scattering data are combed for evidence of either quark or prehadron scattering in the nuclear medium. One curious phenomenon expected to result from this feature of prehadron formation is the so-called "color transparency" of nucleons and mesons as they pass through nuclear material: The nuclear material becomes increasingly more transparent as the momentum imparted to a nucleon or meson increases through the transition from the strongly interacting to the weakly interacting regime. Electron-scattering experiments in which a meson is detected are an excellent way of exploring this feature of QCD. As the momentum imparted to the meson increases, its cross section within the nuclear medium decreases and the medium appears to become transparent. JLAB experiments have yielded evidence for the color transparency effect for mesons produced in nuclei. Another example: If quark-nucleon scattering occurs rather than prehadron nucleus scattering, the detected hadron distributions will be broadened in momentum from the increased interactions with the nuclear

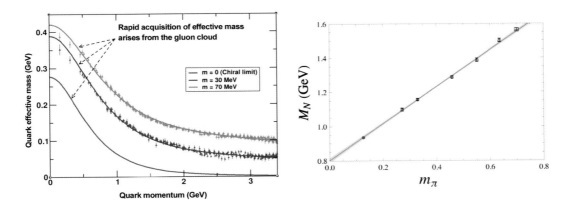

FIGURE 2.32 Mass from nothing. *Left:* In QCD a quark's mass depends on its momentum (in this case it is the magnitude of the "four-momentum," combined momentum and energy). Numerical simulations of lattice QCD (data, at three different bare masses) have confirmed model predictions (solid curves) that the vast bulk of the constituent mass of a light quark is contained in a cloud of gluons, which are dragged along by the quark as it propagates. In this way, a quark that appears to be absolutely massless at high energies (m = 0, red curve) acquires a large constituent mass at low energies, indicated by the leftward rise of the red curve. The other curves show this effect for quarks that are not quite massless at high energies, which makes the numerical lattice QCD calculations tractable. *Right:* Lattice computation of the relationship between the mass of the lightest meson, the pion, and the mass of a nucleon. Nucleons remain massive even if the quark mass, or equivalently the pion mass, is set to zero. SOURCES: *(Left)* Reprinted from C.D. Roberts, Hadron properties and Dyson–Schwinger equations, *Progress in Particle and Nuclear Physics* 61, Copyright 2008, with permission from Elsevier; *(right)* adapted from Fermilab Lattice and MIMD Lattice Computation (MILC) Collaborations. *Physical Review D* 79: 054502 (2009), Figure 10d, Copyright 2009, American Physical Society.

medium. Evidence for such broadening effects is currently being sought using a variety of techniques, including deep inelastic electron scattering in which a meson is produced and Drell-Yan production in which two protons produce two muons. The deep inelastic scattering experiments will be performed at JLAB after its 12 GeV upgrade is complete, and the Drell-Yan experiment is under way at Fermilab, with possible future experiments at RHIC.

Another way to understand the hadronization process is characterizing the amount of energy lost by the quarks as they pass through nuclear material. Developing the picture of both color transparency and hadronization is essential for understanding the behavior of high-energy quarks in ordinary nuclear matter. The empirical description of quark propagation in ordinary nuclei is the baseline against which to compare results from heavy-ion collision experiments at RHIC and the LHC, as described in the section "Exploring Quark-Gluon Plasma." These results show that quark-gluon plasma is remarkably effective at slowing and even stopping high-energy quarks propagating through it.

At the beginning of this century, a new theoretical QCD framework was developed. Known as the Soft-Collinear Effective Theory (SCET), it provides a systematic method for describing processes that produce energetic hadrons and jets, observables allowing an important window onto strong interactions. This theory yields a universal set of simple theoretical tools for handling a wide range of QCD problems in much the same way that nonrelativistic quantum mechanics provides powerful techniques for analyzing many interesting systems in atomic physics. Over the past decade, the study of SCET has allowed nuclear theorists to make discoveries about the nature of QCD, including new approximate symmetries of nature, new techniques for precision analyses, and new treatments of strong interaction nuclear effects that were previously thought to be intractable. Recent examples include (1) the highest precision determination of the strong coupling constant itself, which determines the strength of all QCD interactions, from jets where low-energy QCD plays a key role; (2) improved accuracy for calculations of deep inelastic electron scattering from protons at energies relevant for data from the JLAB upgrade; (3) the categorization of new parameters that describe key QCD effects in heavy B meson decays that are needed to interpret charge-parity symmetry violation studies; and (4) the resolution of discrepancies in the comparison of less accurate theoretical calculations with data on J/$\psi$ distributions. Many promising applications are on the horizon, including the use of SCET to systematically study jet quenching and radiation in heavy-ion collisions at RHIC and the LHC and to study the substructure of hadrons inside jets.

*Can a Nucleus Become Glasslike?*

The electron-proton collider at DESY-Hamburg (HERA) recently showed evidence of a rapid rise in the density of gluons as the momentum of a struck quark decreases inside the proton. In fact, it appeared that this gluon density was continuing to rise even at the lowest quark momenta achievable at DESY, meaning that as experiments gained the ability to see the effects of gluons carrying less and less momentum, they showed that protons can be stuffed with more and more gluons in total. At some point this trend would have to stop and the gluon density should reach saturation. An ingenious technique that might be capable of detecting this saturation is based on the observation that at very high gluon density all mesons, nucleons, and nuclei should appear to be identical. Nuclei in this extreme regime are a form of universal matter, known as "color glass condensate." The density of gluons in a heavy nucleus can far exceed their density in a free proton, so heavy nuclei may be an amplifier for observing the saturation of the gluon density. Colliding electrons and heavy nuclei at high enough energies to get to the highest possible gluon density would be a key element in an experimental program at a future high-energy electron-ion collider.

### How Do the Nucleonic Models Emerge from QCD?

Understanding how the successful low-energy models of nuclear physics based on a nucleonic description emerge from QCD is an exciting and mostly unfinished story. The picture of nuclei as a collection of neutrons and protons has been enormously successful, despite the fact that there is no explicit reference to the quarks and gluons within the neutrons and protons or to the explicit QCD interactions. As discussed in the subsection "Towards a Comprehensive Theory of Nuclei," the common approach has been to use measured nucleon-nucleon interactions to build theoretical frameworks that explain quantitatively the structure of as many nuclei as possible. Before QCD was discovered, it was imagined that this approach would lead to the fundamental theory of nuclear physics. It became evident, though, that if a free nucleon-nucleon interaction is used, then additional forces, like one directly involving three nucleons, are necessary to explain nuclear bound states. This bottom-up approach to nuclear physics has worked well for describing the structure of light nuclei. Other approaches make use of effective nucleon-nucleon interactions, and these have been successful in describing medium- to heavyweight nuclei. Both approaches have benefited from advances in computational power, as these kinds of calculations for realistic systems were impossible in the past (see Figure 2.10). These calculations have allowed nuclear scientists to take enormous strides in developing a quantitative picture of nuclei, with neutrons and protons as the basic building blocks. One approach to systematically tie effective models to QCD, the underlying theory, is to perform lattice QCD calculations for the nucleon-nucleon interaction. During the past decade, work in this direction began, but it is still a daunting challenge. Filling the gap between nucleon degrees of freedom and quark-gluon degrees of freedom, and in so doing understanding how the nucleon picture emerges from QCD, is a major direction for contemporary nuclear theory.

Chiral perturbation theory is an example of an effective theory that goes a long way to using QCD to bridge the gap between the nucleon picture and the microscopic quarks and gluons. Because it incorporates all the appropriate symmetries, chiral perturbation theory is the precise description of how neutrons, protons, and pions interact at low energies in QCD. This description is built through successive approximations, with each better level of approximation allowing the description to be extended to higher energies at the expense of introducing new parameters that must be measured experimentally before the theory can be used to make predictions. Predictions of chiral perturbation theory can be precisely tested in relatively low-energy experiments being carried out at facilities like the High Intensity Gamma Source (HIGS) at Duke University and the gamma-ray beams at Lund, Sweden, and Mainz, Germany. Complementary investigations are also performed at very low momentum transfer as well as neutral pion decay at JLAB. Chiral perturbation theory is a method for applying QCD to low-energy questions involving a few neutrons, protons, or pions even though a quark-by-quark and gluon-by-gluon description cannot be applied.

Lattice calculations provide another important tool with which to understand how QCD describes the evolution from a quark and gluon picture to a picture in which the actors are interacting neutrons and protons. This promising approach is now being pursued throughout the world. Calculations of properties of nucleon-nucleon interactions and the lightest bound nuclei are still very challenging in lattice QCD, because of the large range of energy scales involved. Lattice QCD calculations are performed in a volume with a well-defined lattice spacing. The size of the volume has to be large enough to accommodate the low energy scales involved in nuclear binding, but at the same time the lattice spacing has to be small in order not to distort the high-energy interactions of quarks and gluons. This is a typical example of a multiscale problem with no clear separation of scales, making for a substantial computational challenge. Systematic breakthroughs in computational methods combined with tremendous advances in computational power are making it possible to take on this challenge. Nucleon-nucleon scattering lengths have been computed, albeit with quarks that are heavier than the quarks in QCD. Lattice theorists exploit this strategy and systematically constrain the quark masses toward realistic values, given the available computational power. Preliminary calculations of the binding energies of helium-3 and helium-4 have been computed using this strategy. As new theoretical approaches and algorithms are developed, the artificial world of heavy quarks will evolve into an accurate representation of nuclear physics. These efforts remain a major computational challenge for the next decade.

A new and better understanding of effective field theories (EFT) and the marriage of lattice QCD and chiral perturbation theory offer a reliable approach to comparing QCD to nature. The EFT approach includes a systematic procedure for determining the number of parameters needed to describe the interactions to a certain level of precision. The traditional approach requires measuring these parameters experimentally and then using the EFT to predict other quantities. However, advances in lattice QCD will allow some of these parameters to be calculated from first principles, increasing the predictive power of EFTs for QCD like chiral perturbation theory and increasing the number of experimental observables that can be used to establish with precision the connection between the fundamental building blocks (quarks and gluons) and the physics of light nuclei in QCD.

## Identifying the Full Array of Bound States—
## The Spectroscopy of Mesons and Baryons

Mendeleev's organization of elements into the periodic table of elements in the mid-nineteenth century had a profound impact on the direction of physics. Identification of the systematic patterns of his table and their eventual explanation using quantum mechanics and the quantum theory of electrodynamics are at the foundation of chemistry, much of physical science, and much of modern engineering. The discovery of isotopes and periodic trends in nuclei has similarly been a key development in nuclear physics. The pattern of the isotopes is key, for example, to understand the formation of chemical elements in the interiors of stars and in the evolution of the universe. The identification of the families of leptons and quarks, at present thought to be the most fundamental building blocks of nature, has led to our present understanding of the classification of hadrons and their excited states, but the picture is still incomplete.

One of the most basic applications of QCD is to explain the organization of the masses of hadronic systems: the mesons, the bound states of a quark and an antiquark, and the baryons, which are made up of three valence quarks. QCD should certainly be able to predict the full spectrum. Indeed, hadron spectroscopy experiments in the 1950s and 1960s provided the essential clues that lead to QCD in the 1970s, but some fascinating puzzles remain unsolved. For example, What is the role of gluons in the production of bound states? Why have no states been found with a single well-identified gluon? What is the detailed mechanism that confines quarks within baryons and mesons? What are the most relevant degrees of freedom that explain the experimentally observed spectrum?

Understanding hadron spectroscopy poses many experimental and theoretical challenges. Many excited states are very short-lived and close in energy, making it hard to reliably categorize their quantum numbers or to specify their production mechanism. For almost 50 years the Roper resonance has baffled nuclear physicists. Discovered in 1963 by L. David Roper while working on his Ph.D. at MIT, it is just like the proton, only 50 percent heavier. Its mass was the problem: Until recently, it could not be explained from QCD by any available theoretical method. In a recent breakthrough, theorists at the Excited Baryon Analysis Center (EBAC) at JLAB demonstrated that the Roper resonance is the proton's first radial excitation, with its lower-than-expected mass coming from a quark core shielded by a dense cloud of pions and other mesons. This breakthrough was enabled by both new analysis tools and new high quality data. EBAC has become the Physics Analysis Center, with an expanded scope that includes the analysis of the meson spectrum and their interactions. This is an essential component of the JLAB 12 GeV science program, particularly for the goals of the GlueX experiment in meson spectroscopy.

Identifying the full spectrum of hadrons from first principles with QCD remains a challenge because of the unique and central feature of QCD—namely, that quarks and gluons are confined. Electrons can exist as free or bound particles, and the world is seen by virtue of the existence of freely propagating photons. However, quarks do not exist in nature as free particles and neither do beams of isolated gluons. They are bound within protons, neutrons, pions, and other hadrons. Further complicating the story is the fact that gluons interact with themselves as well as with quarks. This means that there can be QCD bound states made entirely of gluons, with no quarks, aptly dubbed "glueballs." It is also possible for gluons themselves to contribute to the basic properties, such and total angular momentum and parity (i.e., their "quantum numbers"), of bound states with just one quark and one antiquark, resulting in so-called exotic mesons. Experimental searches for these around the world have yielded hints of their existence, but definitive evidence is not yet in hand. Challenges include first providing an environment where production of these states is favorable, and secondly disentangling one potential state from another by uniquely identifying their quantum numbers.

Technical advances have continued at a rapid pace, and there is exceptional potential for progress on the experimental front. Baryon and meson spectroscopy will be a main thrust of the program for the JLAB upgrade, where the CEBAF accelerator will provide beams of polarized gamma-rays. The GlueX experiment at JLAB is being optimized to explore the existence of exotic mesons with a sensitivity that is hundreds of times higher than in previous experiments. Complementary to the exotic light-quark bound states to be studied by GlueX, the Facility for Antiproton and Ion Research (FAIR) at GSI plans to study the heavier exotic charm quark bound states.

Computing the masses of these bound states is a formidable theoretical task. Lattice QCD is a brute force approach to performing these computations. The lattice QCD calculation of the masses of the eight lightest baryons is shown in Figure 2.33, showing remarkably good agreement with experiment. An easier problem is to compute the spectrum of hadrons that contain the heavier charm and bottom quarks on the lattice: the spectrum of bound states with heavy quarks also compares very well with experimental data from the high-energy physics experiments BaBar, Belle, and CLEO. Lattice QCD calculations indicate the existence of yet-undiscovered hadrons such as exotic mesons and glueballs, and it is important to confirm or refute these predictions experimentally to make progress. Progress continues in predicting the spectrum of mesons, both regular and exotic, using state-of-the-art lattice QCD computations, as shown in Figure 2.34. These calculations suggest the presence of many exotic mesons in the region accessible by the GlueX experiment at JLAB, ripe for discovery.

Another computational challenge for the next decade is to determine the energies of the heavier excited states of mesons and baryons. Theorists around the world are cooperating on this task, and the work complements the international

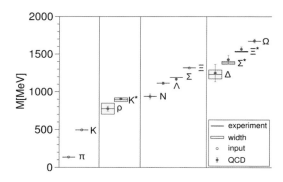 

FIGURE 2.33 *Left:* The masses of the light hadrons, including the lowest lying baryons beginning with protons and neutrons, as predicted by lattice QCD. *Right:* Elements of JLAB's computing facilities for lattice QCD. SOURCE: *(Left)* From S. Dürr, Z. Fodor, J. Frison, C. Heolbling, et al., 2008, *Science* 322: 1224. Reprinted with permission from AAAS; *(right)* courtesy of JLAB.

experimental efforts like GlueX and experiments at FAIR. Progress will come from both significant developments in theory and methodology, but additional investments in state of the art supercomputing facilities will also be required.

Lattice QCD is also being used to understand other aspects of hadronic structure. Calculations of electromagnetic transitions between excited states are another useful connection to experiment. Radiative transitions in charmed mesons have already been calculated, and applications to other systems will come in the next decade. The calculations can be compared with experimental results from CLEO, providing an ideal test bed for the validity of the theoretical approaches used. In the years to come they will be refined and extended to the light quark sector, providing theoretical input to the upcoming experiments.

### Toward the Next Steps: An Electron-Ion Collider

An important next step in nuclear physics will be to connect studies of the quark/gluon structure of nucleons with the study of complex nuclei, by determining how the deep internal structure of nucleons is affected when the nucleons are bound inside nuclei. The nuclear physics community has studied extensively the science that a future electron-ion collider (EIC) could enable with a combination of relatively high center-of-mass energy, high luminosity, and polarized electron beams colliding with beams of polarized protons, light ions, and heavy nuclei. While the technical parameters of an EIC have yet to be finalized with respect to specific science goals, the energy would likely be lower than that of HERA, but the

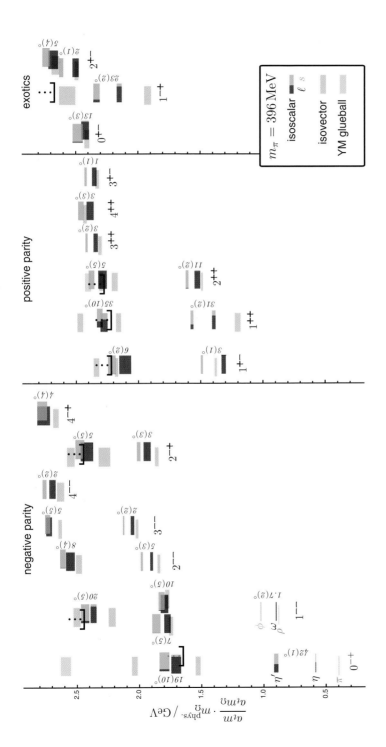

FIGURE 2.34 Results from a lattice QCD calculation of the meson spectrum. The first two panels show the masses of states predicted by the quark model. For comparison the mass of a purely gluonic excitation is plotted in pink. The third panel shows the masses of states that involve excitations of both quarks and gluons and are not predicted by the quark model (exotics). The calculations were performed at unphysical quark masses that result in a pion mass of 390 MeV. However, they suggest the presence of many exotics in the region accessible by the GlueX experiment at JLAB.
SOURCE: J.J. Dudek, R.G. Edwards, B. Joo, M.J. Peardon, D.G. Richards, and C.E. Thomas, 2011, Isoscalar meson spectroscopy from lattice QCD,
*Physical Review D* 83:111502, Figure 4. Copyright 2011, American Physical Society. Courtesy of D.G. Richards, JLAB.

intensity would be as much as 1,000-fold higher and, unlike HERA, the EIC would use nuclear and polarized beams in the collision. Such a capability would provide groundbreaking reach to low momentum partons in the proton and nuclei in the same fashion that the JLAB 12 GeV upgrade can probe the valence quark region of the nucleons and nuclei. An EIC would also provide direct access to the dynamics of the complex system of strongly interacting quarks and gluons that result in the proton's spin. This includes orbital motion, the importance of which research at JLAB and RHIC and within the HERMES experiment has made apparent. Quark orbital motion leads not only to angular momentum but also to significant quark transverse motion in the proton. An EIC would permit access to the transverse distributions, allowing the development of a multidimensional (in space and momentum) image of the sea quarks and gluons. Finally, as mentioned above, an EIC would permit a first serious look at the gluonic structure of the proton and the nucleus, including the remarkable glass-like characteristics expected for the lowest momentum gluons in protons and nuclei.

## FUNDAMENTAL SYMMETRIES

At the end of the nineteenth century, physicist Albert A. Michelson was confident that most of physics had already been discovered, but nevertheless urged still better experimentation:

> The more important fundamental laws and facts of physical science have all been discovered, and these are so firmly established that the possibility of their ever being supplanted in consequence of new discoveries is exceedingly remote. Nevertheless, it has been found that there are apparent exceptions to most of these laws, and this is particularly true when the observations are pushed to a limit, i.e., whenever the circumstances of experiment are such that extreme cases can be examined. Such examination almost surely leads, not to the overthrow of the law, but to the discovery of other facts and laws whose action produces the apparent exceptions.[4]

Michelson vastly underestimated the kind of revolution that would emerge from "pushing observations to a limit." Within a few years, nuclear physics, relativity, and quantum mechanics would change physics forever.

Physicists are now more keenly aware of what remains unknown than at any time in the past, and often formulate the open issues in the form of concise questions. In the nuclear physics of neutrinos and fundamental symmetries, the subject of this section, the questions addressing matter at a very basic level, are these:

- What is the nature of the neutrinos, what are their masses, and how have they shaped the evolution of the cosmos?

---

[4] A.A. Michelson, 1903, *Light Waves and Their Uses,* p. 23, University of Chicago Press.

- Why is there now more visible matter than antimatter in the universe?
- What are the unseen forces that were present in the dawn of the universe but disappeared from view as it evolved? Once very hot and very homogeneous, the universe now displays a preferred "handedness" and so the existence of lost forces.

The predictions of the Standard Model of particles and fields has demonstrated, in some cases to 10-digit precision, our understanding of physics, but it has also helped to place in stark relief that which is yet to be known. The experimental observation of something that is not included in the Standard Model, or is in contradiction to it, is by definition "new physics." It is something that demands the closest attention as it holds the promise of discovery and deeper understanding. As has been known from its beginning some 40 years ago, the Standard Model simply does not speak to certain domains of physics, such as gravity, and in the last decade a definite contradiction to it on its home turf has been demonstrated—namely, the discovery of neutrino mass, in which nuclear physicists played a leading role.

The next decade will find nuclear physicists continuing to look for fingerprints of physics beyond the Standard Model. In addition to exploring the nature of neutrinos, efforts will take place on the precision frontier, where subtle details in the decay patterns of nuclei and the free neutron, in weak interactions between nucleons, and in interactions of electrons in scattering experiments, among others, might signify the presence of new physics.

### A Decade of Discovery

That the mass of neutrinos must be much smaller than that of other matter particles was apparent to Enrico Fermi as he developed the theory of beta decay and compared it to the data available in 1932. By the time the Standard Model was being developed, it was clear that the mass of the neutrinos was so small that the model's mechanism for mass generation would be unnatural for neutrinos. (The most sensitive experiment to date, on the beta decay of tritium, limits the mass to 2.3 eV, already more than 100,000 times smaller than the electron mass.) The only apparent solution was to make the neutrino mass exactly zero and neutrinos purely left-handed. A particle that is always spinning in a left-handed (or right-handed) sense must move at the speed of light and must therefore be massless. Otherwise one could in principle board a fast train and see the particle falling behind, spinning in the opposite direction in its motion relative to the observer. If such particles, moreover, are always found to be left-handed, that is a violation of parity—the concept that physical laws describing a reaction remain the same for both the reaction and its mirrored image. The discovery of parity violation in weak interactions and the measurement of the left-handedness of the neutrino in the 1950s fitted perfectly with the concept of a massless neutrino.

A seemingly unrelated issue was the solar neutrino problem, in which the number of neutrinos detected in the chlorine experiment of Ray Davis, Jr., fell short by a factor of about 3 from theoretical expectations for the solar-neutrino flux tied to energy production in the sun. While many believed this to be due to errors in the theory or the experiment, Bruno Pontecorvo in 1967 raised the possibility that neutrino oscillations might be responsible. Neutrino oscillation is a time-dependent change in the type, or flavor, of a neutrino as it travels, an effect that can only be observed if neutrinos have mass. Flavor transformation and oscillations of neutrinos are possible if the states of well-defined mass (mass eigenstates) are not the same as the flavor eigenstates (the state of neutrinos obtained in beta decay, for example). Davis's detector was sensitive only to the expected flavor, electron, and not to the other flavors, $\mu$ and $\tau$. Hence neutrinos that oscillated to those other flavors would seem to be missing. By the mid-1990s new solar neutrino measurements had been done and the evidence made an astrophysical solution unlikely. Painstaking laboratory measurements of the nuclear reaction rates that determine neutrino production in the sun similarly excluded uncertainties in the nuclear physics of the solar model. On the other hand, neutrinos with mass gave a good account of the observations.

Definitive evidence for neutrino oscillations emerged from another unexpected quarter, the atmospheric neutrino background in proton-decay detectors. Indications from early kiloton-scale detectors that muon neutrinos were about half as abundant as expected were convincingly verified in the 50-kiloton Super-Kamiokande (SK) detector in Japan in 1998. A clear azimuth- and energy-dependent signature of oscillations for neutrinos traversing paths up to the diameter of the earth indicated that muon neutrinos were transforming to an undetected neutrino species. No corresponding effect was seen for electron neutrinos.

Pieces fell rapidly into place in the next few years as the Sudbury Neutrino Observatory (SNO) in Canada, a 1-kiloton heavy water detector for solar neutrinos (see Figure 2.35), showed that electron neutrinos did participate in neutrino oscillations as well. In this case, however, the difference between the squares of the masses is not the same as observed in atmospheric neutrinos; rather it is a factor of about 30 smaller. With three neutrinos there are only two independent mass-squared splittings. SNO showed, furthermore, that the astrophysical theory of the sun was extraordinarily accurate, by correctly predicting the central temperature to a precision of about 1 percent, despite strong skepticism with the solar model expressed beforehand.

The results from SNO and other solar neutrino experiments admit two different solutions for the second splitting. It was not until the KamLAND experiment in Japan detected antineutrinos from distant nuclear power plants fortuitously

FIGURE 2.35 The SNO detector viewed with a fisheye lens. The central acrylic vessel containing 1,000 tons of heavy water ($D_2O$) is 12 m in diameter and surrounded by 9,500 photomultipliers. SOURCE: Courtesy of SNO.

situated with respect to the Kamioka underground site that this ambiguity was resolved in favor of what is termed the large-mixing-angle (LMA) solution. The KamLAND data show clearly the wavelike pattern that is the hallmark of oscillations (see Figure 2.36).

These three experiments have established the basic landscape of neutrino mass and mixing as it is known today. They show conclusively that, in contradiction to the expectation of the minimal Standard Model, neutrinos do have mass, albeit very small (see Figure 1.5). In another very recent, remarkable advance, the third (and last) mixing angle for neutrinos has been measured at the reactor complex at Daya Bay in China by a U.S.-Chinese-Czech-Russian collaboration in which nuclear physicists played a major role. The angle $\theta_{13}$ is a parameter describing how much electron flavor is to be found in the mass eigenstate that is well separated from the other two and is the key ingredient in deducing whether or not neutrinos respect time's arrow. Important questions about neutrinos remain to be answered, as will be described below: What exactly are the masses? Are neutrinos their own

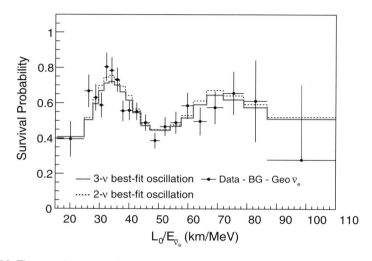

FIGURE 2.36 The neutrino oscillation wavelength is inversely proportional to the neutrino energy $E_\nu$. The oscillation pattern is made strikingly visible in the KamLAND detector by sorting the events according to their energy and using an average distance $L_0$ (about 185 km) to the reactors that produce the (anti)neutrinos. SOURCE: A. Gando, Y. Gando, K. Ichimura, et al. (KamLAND Collaboration), 2011, *Physical Review D* 83: 052002, Figure 5. Copyright 2011, American Physical Society.

antiparticles? Would they behave the same if the arrow of time were reversed, and, if not, did they cause the matter-antimatter asymmetry of the universe?

Through nearly four decades of tests, the Standard Model has otherwise proven to be extremely resilient. But new challenges have emerged besides the neutrino mass. One of the most spectacular achievements of twentieth-century physics is quantum electrodynamics, now a subset of the Standard Model. The magnetic moments of the electron and muon and the Lamb shift in atoms can be theoretically calculated and experimentally measured to precisions of parts per billion or better. As the precision advances, however, corrections from possible contributions that are beyond the Standard Model may become significant. The recent measurement of the anomalous fraction $a$ of the magnetic moment of the muon at BNL gives $116,592,020 \times 10^{-11}$, which is 3.4 standard deviations larger than the theoretical prediction (see Figure 2.37). The anomalous moment of the muon is expected to be sensitive to contributions from new physics, such as are described in the plausible scenario of supersymmetry. The presence of new particles in such theories significantly modifies the effects on the magnetic moment caused by the fleeting appearance and disappearance of charged particles near the muon.

Measuring the Lamb shift (a shift in energy levels specifically predicted by quantum electrodynamics) in muonic hydrogen is a very challenging experiment

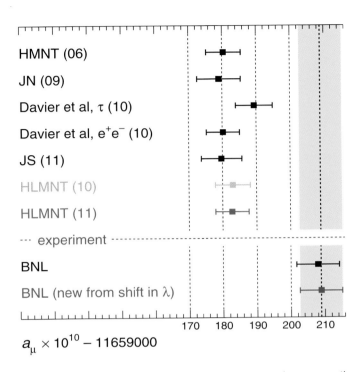

FIGURE 2.37 Comparison of BNL measurement of the muon anomalous magnetic moment (bottommost points) with several recent theoretical calculations based on the Standard Model. HMNT (06), K. Hagiwara, A.D. Martin, D. Nomura, and T. Teubner, 2007, *Physics Letters B* 649: 17; JN (09), F. Jegerlehner and A. Nyffeler, 2009, *Physics Reports* 477: 1; Davier et al., τ(10) and Davier et al., e⁺e⁻ (10), M. Davier, A. Hoecker, B. Malaescu, and Z. Zhang, 2011, *The European Physical Journal C* 71: 1515; JS (11), F. Jegerlehner and R. Szafron, 2011, *The European Physical Journal C* 71: 1632; HLMNT (10), a preliminary version of the work presented in HLMNT (11); HLMNT (11), K. Hagiwara, R. Liao, A.D. Martin, D. Nomura, and T. Teubner, 2011, *Journal of Physics G* 38: 085003; BNL, reported results from Brookhaven National Laboratory. SOURCE: K. Hagiwara, R. Liao, A.D. Martin, D. Nomura, and T. Teubner, 2011, (g − 2)μ and α(M2Z) re-evaluated using new precise data, *Journal of Physics G* 38: 085003. Available online at http://iopscience.iop.org/0954-3899/38/8/085003/. Copyright 2011 IOP Publishing Ltd.

that has recently been carried out for the first time at the Paul Scherrer Institute in Switzerland. In a surprising outcome, the measured shift is more than three standard deviations from the value expected theoretically. It can be interpreted as a discrepant measurement of the radius of the proton, but the cause may lie elsewhere, in experiment or theory, and is not at present known. This surprising result may also, like the anomalous moment, be a hint of physics beyond the Standard Model.

## The Next Steps

Just as quantum electrodynamics was found to be a part of the Standard Model, it is broadly anticipated that the present Standard Model is but a part of a still more comprehensive model. The experimental observations above provide important clues for the discovery of the New Standard Model (NSM), which will incorporate the many successes of the existing model but will in addition provide an understanding of aspects of physics that now are mysterious. What are the dark matter and the dark energy that pervade the universe? Why does the universe contain matter, but little antimatter? What is the origin of the many seemingly arbitrary parameters that emerge in the Standard Model? Are they related and predictable? How can gravity and general relativity be conjoined with the rest of physics? These important questions increasingly call for the knowledge and techniques developed in nuclear physics.

The search for the NSM is proceeding along three complementary frontiers: the high-energy frontier, where experiments at the CERN LHC can discover new particles associated with the NSM; the astrophysical frontier, where measurements of gamma-rays and neutrinos produced in astrophysical environments may uncover the nature of cold dark matter; and a third frontier known in nuclear physics as the precision frontier and in particle physics as the intensity frontier. Among these challenges it is principally the precision frontier—where exquisitely sensitive measurements may reveal tiny deviations from Standard Model predictions and point to the fundamental symmetries of the NSM, or directly reveal the interactions of dark-matter particles and neutrinos—that has attracted the attention and participation of nuclear physicists. Particle physicists are approaching closely related fundamental physics questions with different tools, intense accelerator-produced beams of neutrinos, muons, and kaons.

## The Precision Frontier

As Michelson emphasized and experience has confirmed, it can be very illuminating to subject predictions to the most careful experimental scrutiny possible. In addition to the experiments described above in neutrino physics and muon physics, increasingly stringent precision tests of our present understanding of physics, as embodied in the Standard Model, have been devised. A convenient way to organize them for discussion is by "probe." With few exceptions, they fall into groups: the beta decays of nuclei and the free neutron, weak interactions between nucleons; the weak interactions of electrons; the decay of the muon; and the decay of the pion. Searches for a permanent electric dipole moment (evidence for the violation of the CP symmetry—the product of charge conjugation symmetry (C-symmetry) and parity symmetry (P-symmetry) ) and searches for neutrinoless double-beta decay (evidence that neutrinos and antineutrinos are the same particle) are a particular focus in nuclear physics.

### Beta Decays of Nuclei and the Free Neutron

The beta-decays of nuclei in which both the parent and daughter nuclear states have zero angular momentum and positive parity ("superallowed" nuclear decays) provide a value for the largest and most precise element $V_{ud}$ in the Standard Model Cabibbo-Kobayashi-Maskawa (CKM) matrix that relates quark flavor states to quark mass states. The CKM matrix transforms the quark description with well-defined masses (for which there are no special names) into states with well-defined flavors (down, strange, and bottom). Because there must be a one-to-one relationship between the two descriptions (i.e., no additional quarks beyond the three known to exist), the matrix is unitary, which defines some relationships between the elements and reduces the number of independent parameters to only three, plus a phase that has no effect on the size of the parameters. When combined with the results of kaon and B meson decay studies, which yield the small terms $V_{us}$ and $V_{ub}$, the superallowed nuclear decays provide a stringent test of the unitarity property of the CKM matrix (see Figure 2.38). If this unitarity requirement were found to be violated, it might imply the existence of new interactions such as right-handed weak interactions; an additional generation of quarks and leptons; or the effects of virtual supersymmetric particles that modify the dynamics of the decay. The correlation between the spin axis of a radioactive nucleus and the emission direction of a beta particle or a neutrino can also yield information about possible non-Standard-Model structure of the weak interaction.

Neutrons are neutral particles heavy enough that they are unstable: A free neutron can decay with a half-life of about 10 minutes into a proton, electron, and antineutrino. This decay process makes the neutron a microlaboratory for the study of the weak interaction. When combined with the results of neutron decay correlations, the lifetime of the free neutron provides an independent test of CKM unitarity. The neutron lifetime itself is also one of the key inputs in big bang nucleosynthesis that provides a framework for explaining the abundance of the light elements hydrogen, deuterium, helium-3, helium-4, and lithium-7 in the universe. Notwithstanding its importance, to make a precise measurement of the lifetime at better than the desired one part in a thousand level is very challenging. Improved measurements are needed.

Much more can be learned from a careful study of neutron beta decay. The correlations between the measurable quantities—namely, the neutron spin direction, the emission directions of the electron and the neutrino, the electron spin direction, and the electron energy spectrum—each illuminate a different facet of Standard Model predictions that may disclose the influence of NSM

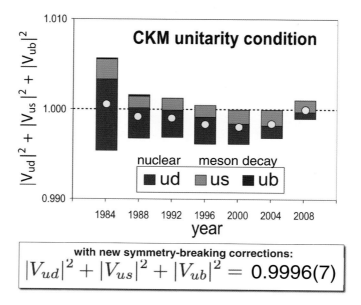

FIGURE 2.38 Experimental evidence supporting the requirement that the Cabibbo-Kobayashi-Maskawa matrix is unitary, as predicted by the Standard Model. The three terms should add up to exactly one, which they do within experimental uncertainty. The colored bars indicate the contributions to the uncertainty from each term, and the yellow dot is the central value. The steady reduction of the uncertainty over the years comes from a worldwide decay-spectroscopy effort, involving rare isotope research at laboratories in the United States, Canada, and Europe, coupled with theoretical advances in calculating the radiative and isospin-breaking corrections. SOURCE: Courtesy of G. Savard, ANL, and J.C. Hardy, Texas A&M University.

physics. A vigorous worldwide program of precise weak decay studies aims to achieve significant improvements in sensitivity. It involves ongoing studies of the superallowed nuclear decays at ANL, Texas A&M University, TRIUMF, Jyvaskyla, ISOLDE, and Munich, and in the future FRIB, where rare, unstable isotopes will provide enhanced sensitivity for testing the theory of correction terms. Improved measurements of the neutron lifetime and neutron decay correlations are planned at the Institut Laue-Langevin, the Los Alamos Neutron Science Center (LANSCE), NIST, TRIUMF, Munich, and the Fundamental Neutron Physics Beamline (FNPB) at the SNS.

*Sterile Neutrinos*

The unitarity of the CKM matrix supports the conclusion that the known quarks are the only ones that exist in nature. In the neutrino world, however, there are intriguing indications that the three known flavors may be accompanied by other, so-called sterile neutrinos that mix slightly with the known neutrinos. Data from the Liquid Scintillator Neutrino Detector at Los Alamos, from a number of short-baseline reactor oscillation experiments, from the MiniBooNE neutrino oscillation search at Fermilab, and from radioactive-source tests of the Soviet-American Gallium Experiment (SAGE) and Gallex/Gallium Neutrino Observatory (GNO) solar neutrino detectors all appear to exhibit small deviations from the three-neutrino expectation. A consistent interpretation has been elusive. The results are limited by statistical and systematic uncertainties, pointing to a need for new tests and theoretical work. The low-energy solar neutrino spectrum remains imprecisely known, and the ongoing measurements by the Borexino experiment in Italy as well as new experiments being designed would permit a comparison between the sun's energy production and its neutrino production. Carried out with sufficient precision, the comparison would be a test of both neutrino unitarity and of our understanding of solar energy generation. Despite the general success of solar models, there are small but significant discrepancies related particularly to the abundance of elements heavier than helium, and new experiments could also provide insight into this problem.

*Weak Interactions Between Nucleons*

The same weak interaction that gives rise to beta decay also contributes to the force between quarks and therefore between nucleons. Its magnitude is tiny ($10^{-14}$) by comparison with the strong force, but it discloses its presence through parity violation because the strong force respects parity. Highly sensitive experiments reveal its presence unequivocally, but one particular part of the weak interaction between nucleons, in which a pion is exchanged, has defied experimental and theoretical quantification. New experiments are under way to try to observe the parity-violating rotation of neutron spin as neutrons pass through matter and a possible preference in spin direction as neutrons are captured by protons. At the same time, advanced lattice-gauge theory is being applied in the hope of achieving a theoretical understanding of the apparent suppression of this part of the force. Theory is currently limited by existing computational resources.

*Weak Interactions of Electrons*

A somewhat complementary avenue involves the measurement of parity-violating (PV) asymmetries in the scattering of longitudinally polarized electrons from nuclei or from other electrons. Historically, the measurement of such an asymmetry in deep inelastic scattering from deuterium at SLAC played a key role in confirming the fundamental prediction of the Standard Model that there were neutral weak interactions. Indeed, two more accurate versions of this classic experiment, the

Parity Violation in Deep Inelastic Scattering (PVDIS) and the PVDIS-SOLID, are planned at JLAB. After the initial work at SLAC, PV electron scattering was used with great success to probe the contributions of strange quarks to the nucleon's electromagnetic properties through a program of measurements at MIT-Bates, the Mainz MAMI facility, and the CEBAF beam at JLAB. A measurement of the PV asymmetry in Moller scattering (in which polarized electrons are scattered from unpolarized ones) performed at SLAC yielded the most precise determination of the dependence of the weak mixing angle on energy scale, one of the more dramatic and novel predictions of the Standard Model. The weak mixing angle, or Weinberg angle, is a parameter of the Standard Model that defines (among other things) the extent to which interactions mediated by the Z boson violate parity. A still more precise version of this experiment, Moller, is planned at JLAB after the completion of the energy upgrade. It would complement another PV experiment, Q-weak, currently under way at JLAB involving elastic scattering from a proton target. Together, the comparison of results of purely leptonic (Moller scattering) and semileptonic (electron-proton scattering) experiments can provide a powerful test of the Standard Model.

*Muon Decay*

The properties and decays of muons—structureless particles like electrons but with a mass about 200 times greater—are among the most sensitive probes of the Standard Model. The example of the muon anomalous moment was already described above. In addition, nuclear physicists have recently reported new results on the correlation parameters in muon decay, the muon lifetime, and muon capture in hydrogen, which have improved previous experimental values by factors of 10 or more. The muon lifetime, now determined to part-per-million accuracy, defines the strength of the weak interaction. These new results give the tightest limits now available on interactions beyond those in the Standard Model. Over the next decade, new measurements with muons will continue to push the precision frontier. An experiment to search for the decay of a muon into an electron and photon with a 100-fold better sensitivity than previous measurements is under way now at the Paul Scherrer Institute (PSI). This conversion is essentially forbidden in the Standard Model but is predicted to occur in certain proposed theoretical extensions. Two experiments—a new, even more precise measurement of the anomalous moment of the muon and a sensitive search for the conversion of a muon to an electron in the field of a nucleus—are being planned by collaborations of high-energy and nuclear physicists for Fermilab following its intensity upgrade.

*Pion Decay*

Like the neutron, the pion is a composite particle that undergoes beta decay, but because of its large mass, it can decay to either an electron or a muon with an associated neutrino. The relative decay probabilities of charged pions into a muon or an electron have provided stringent (better than 0.1 percent) tests of the lepton universality property of the Standard Model weak interaction, which simply states that the weak interaction acts with the same strength in every family of elementary particles. With the operation of the LHC and its prospective sensitivity to the existence of new particles with multi-TeV masses, the forefront sensitivity for many of the weak decay studies will rise to 1 part in 10,000 within the next decade. Two new pion beta decay measurements are planned at TRIUMF and PSI.

## Two Challenges

Certain experimental efforts in nuclear physics are motivated by specific expectations for the physics that the NSM is likely to display. Two specific research thrusts having great discovery potential are searches for the permanent electric dipole moments (EDMs) of the nucleon, neutral atoms, and charged leptons and searches for the neutrinoless double beta decay of heavy nuclei. Hand in hand with these experimental initiatives is a focused program of theoretical nuclear physics studies that aim to interpret the results of these and other experiments in terms of the NSM.

*Search for a Permanent Electric Dipole Moment*

The goal of the EDM searches is to discover a mechanism for the violation of CP symmetry (or time-reversal symmetry) beyond the CP violation that can be accounted for by the Standard Model weak interactions. The reason is that an explanation of the excess of matter over antimatter in the present universe requires the existence of a not-yet-understood source of CP violation in the early universe. Perhaps it may be found in the neutrinos, as we consider below. Alternatively, if the matter-antimatter asymmetry was produced when the universe was roughly 10 picoseconds old—during the era of so-called electroweak symmetry breaking—then the next generation of EDM experiments would have a good chance of observing it. EDM searches look for a small shift in the precision frequency of a quantum system with spin (such as the neutron) in the presence of electric and magnetic fields. An EDM violates both parity (P) and time-reversal (T) symmetry, but not the matter-antimatter symmetry C. In addition to uncovering the CP violation needed to explain the matter-antimatter asymmetry, the EDM searches could also reveal the presence of CP violation in the strong interaction. The present limits on the latter are so stringent as to imply the possible existence of another symmetry, known as Peccei-Quinn symmetry, a symmetry invoked specifically to explain why the CP symmetry is not violated in the strong interactions at more than about $10^{-10}$. The violation of this symmetry in such a way as to lead to a nonvanishing EDM would imply the existence of a new particle called the axion. If it exists, the axion itself could also make up the cosmic dark matter.

The next generation of EDM searches is expected to improve the level of sensitivity by up to two orders of magnitude over present limits. Intensive efforts to reach this level of sensitivity are under way in the United States, Canada, and Europe. They include searches for (1) the neutron EDM at the Fundamental Neutron Physics Beamline at the Oak Ridge SNS, the Institut Laue-Langevin in Grenoble, and the Paul Scherrer Institute in Switzerland; (2) the atomic EDMs of mercury, radium, radon, and xenon at various laboratories and universities; and (3) the EDM of the electron, using molecular or solid-state systems, in the United States and Europe. In addition, nuclear scientists at BNL are developing a possible measurement of the proton EDM using a storage-ring technique. The "physics reach" of these searches expressed in terms of the mass of new, presently unknown particles is in many cases at a scale beyond that accessible at the LHC. The LHC gets its sensitivity by directly trying to produce and detect new particles involved in CP-violating interactions, while the experiments that are the main subject of this paragraph look for the effects of the same interactions at much lower energies by seeking rare effects induced by quantum fluctuations. The present EDM limits generically imply that the mass scale of any new CP-violating interaction (in other words, the mass of some new particle that could mediate the interaction) is greater than several TeV, and improvements by two orders of magnitude would extend this scale by a factor of 10, well beyond the scale accessible at the energy frontier.

### Search for Neutrinoless Double-Beta Decay

Because neutrinos lack electric charge they can in principle be their own antiparticles. Whether some symmetry preserves a distinction between matter and antimatter for neutrinos is presently unknown. The answer to this question may be at the heart of why the universe contains matter and essentially no antimatter, because the violation of total lepton number could be associated with the generation of the matter-antimatter asymmetry at times much earlier than 10 picoseconds after the big bang. It is also a question that needs an answer for the construction of the NSM, because it leads to a novel mechanism for the generation of particle mass, one that does not exist in the Standard Model.

The only practical experimental approach to this problem is the search for neutrinoless double beta decay. The pairing property of the nuclear force leads to a number of nuclei that are stable against all decay modes except the simultaneous emission of two electrons and two antineutrinos. The process, while allowed, rarely occurs. Of approximately 10 examples known, the shortest half-life is still a billion times longer than the age of the universe. If neutrinos and antineutrinos are the same particle, then the decay can proceed with the emission of just the two electrons and no neutrinos—that is, neutrinoless double beta decay. That process has not yet been seen, with lifetime limits some $10^4$ times longer still than the two-neutrino mode. A major experimental attack on this problem, calling ultimately for detectors containing a ton or more of an enriched isotope, is a priority in nuclear physics. In addition to answering the question of whether a neutrino is the same as an antineutrino, a Majorana particle, or is different, a Dirac particle, a positive observation would help to define the mass of neutrinos.

As with the EDM experiments, there exists a worldwide program of searches for neutrinoless double beta decay. U.S. nuclear scientists are involved in several of these efforts, including the CUORE experiment at Gran Sasso, the EXO experiment at the Waste Isolation Pilot Plant (WIPP) in New Mexico, the Majorana Demonstrator Project at the Sanford Underground Laboratory, SNO+ at SNOLAB, and KamLAND-Xen at Kamioka. Majorana neutrinos with masses in the presently allowed range may produce a signal in these experiments. If necessary, larger and more ambitious experiments using enriched isotopes could improve the sensitivity substantially. A next-generation ton-scale neutrinoless double beta decay experiment could be carried out 7,400 feet down in the Sanford Underground Research Facility (SURF) in the Homestake mine in Lead, South Dakota.

### Nuclear Theory at the Precision Frontier

For these experimental efforts at the precision frontier, input and guidance from nuclear theory is vital. For example, interpreting the results of EDM searches in terms of a new mechanism for CP violation and relating the latter to the cosmic matter-antimatter asymmetry requires a web of nuclear theory computations along with calculations from cosmology and astrophysics. Starting from the computation of low-energy matrix elements in strongly interacting systems such as the neutron or mercury nucleus, one must then derive values for the parameters of an underlying model at the elementary particle level, taking into account the constraints from studies at the high-energy and astrophysical frontiers. Computations of the matter-antimatter asymmetry require calculations analogous to those performed when interpreting the results of relativistic heavy ion collisions. A similar chain of theoretical analyses is needed to interpret the neutrinoless double-beta decay results, as well as those from weak decays and PV electron scattering, in terms of the structure of the NSM. The increasing scope of the experimental effort in this area of nuclear science calls for concomitant increases in the related theoretical effort as well as advances in computational tools.

*Connections with Cosmology*

Much remains to be understood about neutrino mass and mixing, and new experiments are under construction or in operation. Oscillations set a lower limit on the mass, but other techniques are required to determine the actual magnitude of neutrino mass. The mass of the lightest neutrino cannot be less than zero, and it follows from oscillation data that the sum of the three masses must be at least 0.06 eV. An upper limit that is independent of assumptions about the properties of neutrinos comes from laboratory measurements of the shape of beta spectra near the end point. There the electron energy approaches the maximum value for the decay, which is limited by the rest mass of the accompanying neutrino. Experimental measurements of the shape of the tritium beta spectrum yield an upper limit on the sum of the three masses of 6 eV, setting a range of 0.06 to 6 eV in which the mass sum must lie. A new, large-scale tritium experiment, the KArlsruhe TRItium Neutrino (KATRIN) experiment, is under construction that will have 0.6 eV sensitivity. New ideas for extending the sensitivity of beta experiments are being explored should the mass sum turn out to be smaller than 0.6 eV.

There is a strong prediction, but not yet direct experimental proof, of a cosmological relic neutrino background. The energies of these neutrinos are so low that detecting them appears all but impossible. However, cosmological arguments also relate the large-scale structure in the universe to neutrino mass. These arguments are model-dependent, being sensitive to the equation of state of dark energy and to the power spectral index that describes how quantum fluctuations in the big bang were distributed in scale. For reasonable assumptions, they limit the mass sum to about 0.6 eV or less. The ESA Planck satellite, launched in 2009, together with new galaxy surveys, may be able to extend the sensitivity to about 0.1 eV. A laboratory measurement at this level would be the most direct laboratory confirmation of the existence of the relic neutrino background that can presently be envisaged and would subject cosmological models to an important test.

Overwhelming evidence from observational astronomy for the existence of dark matter demands an understanding of its particle nature. Neutrinos are now known to be insufficiently massive, and no other known Standard Model particle can explain the data. Many candidates have been advanced, of which two are strongly motivated by theoretical considerations outside of astronomy. The lightest neutral particle in theories such as supersymmetry would be long-lived or stable and could have the mass (still many times the proton mass) and the interaction cross section to be the dark matter. Alternatively, a new symmetry would explain why CP is so well conserved in the strong interactions and would imply the existence of a very light, long-lived particle, the axion, which could also be the dark matter.

Detection of the former type of particle, the weakly interacting massive particle (WIMP), might be achieved by observing the recoil energy imparted to a nucleus struck by a WIMP present in the galactic dark-matter cloud. The energies are small, the interactions are rare, and the backgrounds present significant challenges, but there has been steady progress toward achieving the necessary sensitivity and redundant criteria for identification. Nuclear physics techniques are widely used in this field, and nuclear physicists have much to contribute; indeed, there is great enthusiasm in the nuclear physics community for addressing this challenge.

## Underground Science

Key parts of the experimental program in fundamental symmetries and neutrino astrophysics demand an underground location shielded from the steady rain of cosmic rays that arrive at Earth's surface. The signals from solar neutrinos, supernova neutrinos, and geoneutrinos, from neutrinoless double-beta decay, and from dark-matter particles are so rare that the cosmic ray background at Earth's surface overwhelms them. Deep underground, the flux of energetic muons, the most penetrating cosmic ray particle other than neutrinos, decreases by a factor of about 10 for each 300 m.

The deepest underground research laboratory today is SNOLAB in Canada, where the SNO experiment was carried out at a depth of 2,000 m. A smaller but even deeper laboratory is being commissioned at Jinping in China. Many countries have deep underground research laboratories: the Gran Sasso National Laboratory in Italy at an effective depth of 1,300 m is the largest in the world. In the United States, Ray Davis's experiment on solar neutrinos, for which he shared the 2002 Nobel prize, was carried out 1,600-m down in the Homestake gold mine in South Dakota. Other research sites in the United States include WIPP and the Soudan mine in Minnesota, where neutrinos from Fermilab are detected. Both are about 700 m deep and are confirming the atmospheric neutrino signal and providing increasingly precise data on the "atmospheric" mass-squared splitting.

The priority of the research goals that need underground space, including neutrinoless double beta decay, dark matter searches, and solar neutrino physics, prompted the National Science Foundation to solicit proposals for a science program and a laboratory. Eight sites were proposed, and the Homestake mine was selected for the final design and facility proposal. The owners of the mine had decided in 2000 to terminate commercial operations there. Once closed, the mine flooded and required rehabilitation. The importance of the science and its location in South Dakota attracted private funding in excess of $70 million, unprecedented in the field of nuclear and particle physics. That funding, with additional support from the state of South Dakota, was used to prepare surface facilities for research and to rehabilitate the mine to a depth of 1,300 m. In the interim, the field of high-energy physics became increasingly interested in this area of research and developed plans for a neutrino beam that would originate at Fermilab, 1,300 km to the East. At Homestake, a very large detector would make possible studies of the neutrino mass hierarchy (that is, the ordering of the mass eigenstates by increasing mass), the possible violation of the CP symmetry in neutrinos, and searches for proton decay. If CP violation is observable among neutrinos, it is a tantalizing possibility for explaining why the universe contains mostly matter and not much antimatter. There are both logistical and intellectual advantages for nuclear and particle physicists to collocate in the Sanford Underground Research Facility at Homestake.[5] However, at the time this report was being prepared, the future of that facility and the experiments planned for it were uncertain.

## Fundamental Symmetries Studies in the United States and Around the World

The field of fundamental symmetries is a microcosm for some of the difficulties encountered in managing science in the United States and elsewhere because it does not always fit comfortably within the mainstream of a field. The questions often call for exploration of physics that lies at the interface between two or more disciplines. The physics outcomes are often highly uncertain when a project is starting up.

Many nations have elected to organize a separate research field that, broadly speaking, encompasses particle and nuclear physics, high-energy astrophysics, and cosmology. In the United States, the core disciplines of nuclear physics, particle physics, astronomy, and space sciences have been preserved at the federal agency level and are the homes for investigations in the interface areas often explored in the area of fundamental symmetries. Agency decisions on which discipline area will consider funding an investigation may appear to be arbitrary, but there has been a commendable effort to be flexible and to prevent research from falling into the cracks. Nevertheless, in the competition for scarce resources, core studies in a particular discipline area are likely to enjoy the home-arena advantage when competing against studies that might arguably belong to another discipline. The European approach, forming a separate discipline area, is one solution, but there is also merit in the continuous competition between research at the core of a discipline and research at its boundaries. From such competition, the center of a discipline can begin to shift.

---

[5] NRC, 2012, *An Assessment of the Science Proposed for the Deep Underground Science and Engineering Laboratory (DUSEL)*, Washington, D.C.: The National Academies Press.

## The Workforce

The field of fundamental symmetries and neutrinos serves as a magnet for attracting new talent into physics and its related disciplines. Scientists young and old find the questions at once grand and simple. Motivation does not need to be accompanied by specialized knowledge at the beginning. University physics faculties as well are enthusiastic about the potential of the field for discovery and about the fact that the basic concepts are easily communicated to students and to colleagues. As a result, departments have hired new faculty working in this field. The experimental tools and expertise lie at the field's boundary with particle, atomic, and molecular physics. For students, the field provides exposure to a variety of experimental and theoretical techniques and the opportunity to work at the interface of several disciplines. That breadth of experience is attractive to future employers whether in academe, at the national laboratories, or in industry.

# Highlight:
# Diagnosing Cancer with Positron Emission Tomography

Atomic nuclei with fewer neutrons than stable isotopes decay predominantly by emitting a positively charged electron, a positron, which is annihilated with electrons, emitting gamma-radiation. For more than 35 years positron emission tomography (PET) has been used as a research tool in neuroscience and in diagnosing cancer.[1]

PET imaging makes use of the self-collimating nature of positron decay (see Figure PET 1) as two nearly collinear photons are used to locate an annihilation event. PET cameras are typically made of a ring(s) of detectors that are in timed coincidence (resolving time of a few nanoseconds), allowing a line of response to define the chord along which the positron was annihilated (the location of the emission is not known because of the short distance the positron travels before annihilation, typically a few millimeters). By mathematically back-projecting the lines of response, a density map can be generated that reflects the distribution of the positron emitter.

Functional imaging using PET started as a research tool in neuroscience in the late 1970s and remains a major research tool for the neurosciences. However, its main impact recently has been in the diagnosis of cancer. Originally, simple tracer molecules such as water, carbon monoxide, and carbon dioxide were used. The first complex molecule to be used extensively was the glucose analog $^{18}$F-fluorodeoxyglucose (FDG), developed at BNL in collaboration

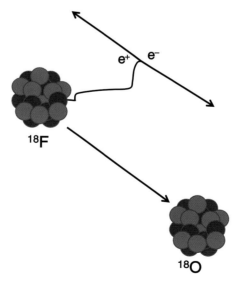

FIGURE PET 1  Illustration of positron decay. One of the protons (red) in the unstable nucleus is converted into a neutron (blue) with the emission of a positive electron (positron). The positron travels a short distance until it is annihilated with a neighboring atomic electron, resulting in two photons (Ð-rays), each with an energy of 511 keV. The photons will travel at nearly 180° from each other to conserve momentum. SOURCE: T. Ruth, 2011, The uses of radiotracers in the life sciences, *Reports on Progress in Physics* 72: 016701. Printed with permission from IOP Publishing Ltd.

with researchers at the NIH and the University of Pennsylvania around 1975. Since the human brain uses glucose as its primary energy source, the availability of the tracer led to groundbreaking studies of the human brain in health and disease. This effort was driven by the successful use of $^{14}$C-labeled deoxyglucose at the NIH by Louis Sokolov in the 1960s. Since $^{14}$C is not detectable from outside of the body, the effort went into developing a labeled analog that could be shipped from a cyclotron facility (BNL in this case) and the PET camera (the University of Pennsylvania). Thus $^{18}$F with its nearly 2-hour half-life became the radionuclide of choice.

Many more tracers are used to investigate the various neuronal systems, probing both the presynaptic and postsynaptic pathways. Several hundred tracers have been prepared and tested for their utility in investigating various enzymatic and receptor systems although only a handful are routinely used. There are tracers specifically designed to monitor cell proliferation, the hypoxic nature of cells, and cell apoptosis.

The heart of the PET camera is the detection system, which builds on scintillator detector systems developed by nuclear scientists. The vast majority of modern PET scanners make use of segmented inorganic scintillation crystals coupled to multiple photomultiplier tubes (PMT). The ideal crystal will have a high stopping power for the 511-keV annihilation photons (high photoelectric absorption), a high light output with wavelength matched to the PMT, and a fast decay time for the light, and it will be physically robust. For nearly two decades the detector material of choice was bismuth orthogermanate, a scintillator often used in basic nuclear science research. More recently, lutetium orthosilicate was introduced. Owing to its higher light output, the segmentation of the crystals could be finer, thus reducing the crystal element size from approximately (4 mm × 4 mm) to (2 mm × 2 mm). There are proposals to reduce the crystal elements to below 1 mm$^2$. In order to accomplish such a task, the packing fraction of the crystals must be improved—in other words, the empty space between crystal elements must remain a small fraction of the total area.

The typical crystal is segmented into an 8 × 8 grid (or more) coupled to four PMTs. There is an algorithm to identify the location of the event by comparing the light sharing among the PMTs. While this scheme reduces the cost of the scanner, there is a loss in resolution owing to the approximate nature of the light-sharing approach. There are prototype scanners using avalanche photodiodes coupled to individual crystal elements, making the finer pixel identification better. Thus far such systems have been built only for small animal scanners.

As the physical limitations of detection are approached, the remaining avenue is to increase the signal to noise ratio by utilizing tracers that are uniquely suited to imaging the function in question and that otherwise clear rapidly from surrounding tissue. To this end, the development of more specific tracers is believed to be the most critical issue for PET.

One of the main strengths of PET compared to single-photon emission computed tomography (SPECT) is the ability to measure, directly, the attenuation effect of the object being viewed. This is the result of requiring that both photons be detected. Thus, if one photon of the pair is not observed, then there is no line of response. Along the path to the detectors, one or both photons (511 keV each, the rest mass of the electron) can undergo absorption. Thus, in order to be detected as an event, both photons must be detected in temporal coincidence. By using an external source of positron emitter, the attenuating (absorbing) extent of the object to be measured can be determined. All commercial PET cameras are now built with a CT scanner (X-ray tomography) so that a merged image of structure and function can be obtained. Since the CT image is a measure of electron density, it is used to calculate the necessary coefficients for attenuation correction. The primary function of the CT image is to provide a detailed view of the section of the body under investigation. Figure PET 2 illustrates the power of this approach.

There are several physical limitations inherent in PET technology. First, as the emitted positron has kinetic energy, varying from a few hundred keV to several MeV depending on

*continued*

which radionuclide is used, it will travel a few millimeters to centimeters before annihilating with an atomic electron. As such, the site of annihilation is not the site of emission, resulting in a limitation when defining the origin of the decay. Another limitation is the fact that the positron-electron pair is not at rest when the annihilation occurs; thus by virtue of the conserva-

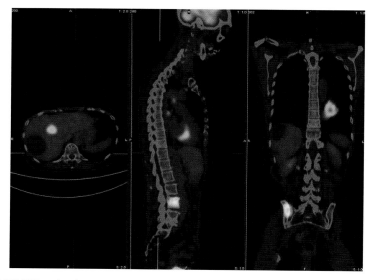

FIGURE PET 2  The three panels from left to right show a combined FDG PET/computed tomography (CT) image in transaxial, saggital, and coronal views. The colored hot metal image is the PET image and the gray image is from the CT camera. The combined image enables physicians to determine the precise location of abnormal function (high uptake in the mass visible on the chest wall in the CT image in this case). In addition, a metastatic tumor is visible in the pelvic region. SOURCE: T. Ruth, 2011, The uses of radiotracers in the life sciences, *Reports on Progress in Physics* 72: 016701. Photo courtesy of British Columbia Cancer Agency and reprinted with permission from IOP Publishing Ltd.

tion of momentum, the two photons are not exactly collinear. Although the lack of collinearity becomes increasingly important with greater detector separation, this effect is ignored, for the most part, in existing tomographs because the detector ring diameter is less than a meter, at which distance the deviation from 180° is a fraction of a millimeter.

Because diagnostic imaging is driven by a digital approach (present/absent, yes/no), the desire to have uncluttered images resulting from PET is very important. Nevertheless, the true power of PET lies in its ability to track the distribution of a tracer over time and to extract detailed kinetic data, as in a physical chemistry experiment where rate constants are determined. So, the conflict between using PET technology for clinical diagnosis and using it as an in vivo biochemistry tool will not be easily resolved, nor should it be.

---

[1]The information in this vignette is adapted from T.J. Ruth, 2009, The uses of radiotracers in the life sciences, *Report on Progress in Physics* 72: 016701. Permission granted by IOP Publishing, Ltd.

# 3

# Societal Applications and Benefits

Nuclear physics is ubiquitous in our lives: Detecting smoke in our homes, testing for and treating cancer, and monitoring cargo for contraband are just some of the ways that nuclear physics and the techniques it has spawned make a difference in our safety, health, and security. Many of today's most important advancements in medicine, materials, energy, security, climatology, and dozens of other sciences emanate from the wellspring of basic research and development in nuclear physics. Answers to some of the most important questions facing our planet will come from nuclear science, interdisciplinary efforts in energy and climate, and marketplace innovations. The economic impact of the applications of nuclear physics is significant. As an example, particle beams from accelerators are used to process, treat or inspect a wide range of products with a collective value of more than $500 billion.[1] At the same time, approximately 23 million nuclear medicine procedures are carried out each year in the United States to diagnose and treat cancers, cardiovascular disease, and certain neurological disorders. In the future, basic nuclear science will be a key discipline that provides ideas and insights leading to the intellectual properties and patents with which venture capitalists and entrepreneurs will shape the economies of the future.

Between the chapters of this document the committee has highlighted some of the ways that nuclear physics impacts our lives along with some of the individuals poised for leadership in nuclear physics. In this chapter we provide a more detailed

---

[1] Department of Energy, 2010, *Accelerators for America's Future,* Washington, D.C. Available at http://www.acceleratorsamerica.org/report/index.html; last accessed August 31, 2011.

overview of some of the ways in which nuclear physics is being applied to address the nation's challenges in health, homeland and national security, nuclear energy, and some of the innovations taking place in developing and exploiting new technologies arising from nuclear science.

## DIAGNOSING AND CURING MEDICAL CONDITIONS

Nuclear physics techniques have been revolutionary in medical diagnostics and cancer therapy. Of the 23 million nuclear medicine imaging and therapeutic procedures performed each year in the United States, typically 40-50 percent are for cardiac applications, while 25-40 percent are for cancer identification and therapy. In addition, nuclear medicine procedures are used to diagnose Alzheimer's disease, treat hyperthyroidism, assess coronary artery disease, localize tumors, and diagnose pulmonary emboli.

The science of nuclear medicine, however, goes far beyond the radiopharmaceuticals used for imaging and treatment. Advances in the field are inevitably tied to basic research in nuclear physics at all levels. These advances include accelerators, detectors, understanding the interaction of radiation with matter, and creating complex statistical algorithms for identifying relevant data.

### Nuclear Imaging of Disease and Functions

Over the past few decades, new nuclear imaging technologies have enhanced the effectiveness of health care and enabled physicians to diagnose different types of cancers, cardiovascular diseases, and neurological disorders in their early stages. Today there are over 100 nuclear imaging procedures available. These procedures have the additional advantage of being noninvasive alternatives to biopsy or surgery. Unlike other imaging procedures that are designed mainly to identify structure, nuclear medicine can also provide information about the function of virtually every major organ system within the body.

The most important modern advances in nuclear imaging are positron emission tomography (PET) and single-photon emission computed tomography (SPECT). PET, especially when coupled to X-ray computed tomography (CT) scans, has become a highly sensitive probe of abnormal functions, as described in detail in the PET highlight between Chapters 2 and 3.

$^{18}$F-fluorodeoxyglucose (18F-FDG) is a radiopharmaceutical used in medical-imaging PET scans. This is a glucose analog that is absorbed by cells such as those in the brain and kidneys as well as cancer cells, which use high amounts of glucose. This procedure yields scans such as those displayed in Figure 3.1 and can be used for the study of organ functions and, in the case of cancer cells, for therapeutic applications. The 1.8-hour half-life $t_{1/2}$ of fluorine-18 results in very high specific

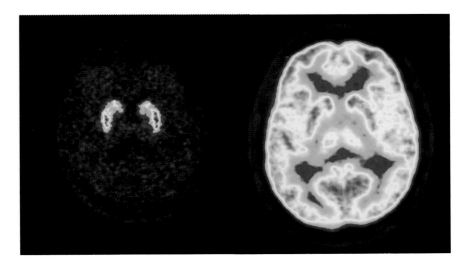

FIGURE 3.1 PET is a powerful tool to probe the functions of the brain. In these images of the brain, the radionuclide is fluorine-18 while the molecules for each image obviously have different biodistributions. The left-hand figure shows fluorodopa (to probe dopamine integrity) while the right-hand figure shows fluorodeoxyglucose (to probe sugar metabolism). SOURCE: Courtesy of Don Wilson, British Columbia Cancer Agency.

activity with no long-term residual activity in the body. However, the short lifetime means that fluorine-18 and 18F-FDG have to be produced very near to where the procedures are to be performed. This often requires in situ small-scale particle accelerators, another capability developed by nuclear physicists, to produce the isotope.

Radionuclides that emit gamma-rays have a long history as imaging tools in the diagnosis of cancer. SPECT has been built around the gamma-ray associated with the decay of molybdenum-99. Molybdenum-99 decays ($t_{1/2}$ = 66 hours) into an isomer of technetium-99m (m indicating metastable), which in turn decays ($t_{1/2}$ = 6 hours) by emitting a 140-keV gamma-ray. The cameras for this imaging technique are typically made with a cluster of photomultipliers coupled to a large NaI crystal. In recent years, the semiconductor material CdZnTe (CZT) has gained favor because of its higher energy resolution. Having this type of capability means that multiple tracers can be imaged simultaneously through the use of different energy windows.

In North America, the main radioisotopes needed for imaging and treatment are produced by the Isotope Development & Production for Research and Applications (IDPRA) program, in the Nuclear Physics Program of the Department of

Energy's (DOE's) Office of Science, and by two Canadian facilities, TRIUMF and Chalk River.

Worldwide, the molybdenum-99/technetium-99m radionuclide pair is used in four out of five, or in about 12 million diagnostic-imaging procedures in nuclear medicine every year. However, the reactors that have been producing molybdenum-99 are approaching the end of their useful lives, which is expected to trigger an "isotope crisis." One of the reactors, the Canadian National Research Universal (NRU) reactor at Chalk River, is scheduled to stop isotope production in 2016, while potential replacement reactors around the world may not be available until 2020. Research is now focused on exploring accelerator-based production of molybdenum-99 as an alternative technology using, among other reactions, the $^{100}$Mo(g,n)$^{99}$Mo and the $^{100}$Mo(p,2n)$^{99m}$Tc reactions.

Another option centers on rhenium-186, which has a favorable half-life ($t_{1/2} = $ 90 hours) and emits beta decay electrons of 0.9 MeV with a 10 percent branch emitting a gamma-ray with energy similar to that of technetium-99m. Since rhenium is in the same chemical family as technetium, much of the technology developed for technetium-99m can be applied to rhenium-186. Current efforts are concentrated on reactor production of rhenium-186 via the $^{185}$Re(n,g) reaction, followed by mass separation to yield a sample with the high specific activity needed for therapy (see Box 3.1).

## New Radioisotopes for Targeted Radioimmunotherapy

Radiopharmaceuticals have been developed that can be targeted directly at the organ being treated. These therapy radiopharmaceuticals rely on the destructive power of ionizing radiation at short ranges, which minimizes damage to neighboring organs.

A frontier direction is targeted radiopharmaceuticals. This involves attaching a relatively short-lived radioactive isotope that decays via high-energy transfer radiation (alpha-particle emission, for example) to a biologically active molecule, like a monoclonal antibody that has a high affinity for binding to receptors on cancer tumors. When the radioactive nuclei decay, the radiation they produce loses energy quickly and because it does not travel far, a lethal dose of radiation is delivered only to adjoining tumor cells. By careful construction of the targeting molecule, the radioactive nuclei will pass through the body quickly if they do not bind to tumor cells, thus minimizing the exposure of healthy tissue to the high-energy transfer radiation. Presently, the most common radionuclides are iodine-131 and yttrium-90, though neither is ideal. Two radiopharmaceuticals, Bexxar (using iodine-131) and Zevalin (using indium-111 or yttrium-90), are now in use to treat non-Hodgkins lymphoma.

Many research efforts are focused on the production of alternative isotopes

## BOX 3.1
## Suzanne Lapi and Radionuclide Production

Suzanne Lapi is a leader in the effort to develop rhenium-186 for radiation therapy. After receiving her master of science and Ph.D. degrees from Simon Fraser University, British Columbia, she pursued research into the production of rhenium-186 of high specific activity  to enhance the therapeutic efficacy of this promising radionuclide. After concluding that accelerator production was not optimal, she focused on increasing specific activity of rhenium-186, produced in a reactor by the $^{185}Re(n,\gamma)$ reaction, by mass separation of the postirradiated material. This work is the subject of a patent and is also being applied to increasing the specific activity of molybdenum-99, also produced via the $(n,\gamma)$ reaction. Presently Dr. Lapi is an assistant professor at the Mallinckrodt Institute of Radiology at Washington University in St. Louis, Missouri. She is a project leader on radionuclide research for cancer applications, oversees production of nonstandard PET radionuclides, and collaborates with internal and external faculty on grants supported by both DOE and the National Institutes of Health (NIH).

FIGURE 3.1.1 Suzanne Lapi. Source Photo courtesy of MIR Photography

with superior cytotoxicity for use in therapy. A promising class of isotopes is those that decay by alpha emission, since alpha particles have a very short range in tissue, resulting in an enhanced cytotoxicity. The radionuclide actinium-225 combines several favorable properties, including a half-life of 10 days, high alpha-particle energy, versatile coordination chemistry, and several alpha-emitting daughter isotopes. Actinium-225 has been used in Phase I and II clinical trials; it is presently being produced at Oak Ridge National Laboratory (ORNL) and at the Institute for Transuranium Elements in Karlsruhe, Germany. Its availability, however, is currently limited, and alternative production mechanisms are being investigated at the Los Alamos National Laboratory (LANL) isotope production facility. More recently, researchers at the Karlsruhe Institute in Germany have reported the efficacy of treating neuroendocrine tumors with the alpha-emitting bismuth-213 nucleus attached to a biological molecule (called DOTATOC) that targets these particular tumors. They found that the tumors of seven out of nine patients had become smaller with no discernible negative side effects. If this approach can be validated and brought into routine use, the treatment of cancer will have had a major paradigm shift.

In the coming decade, nuclear physics facilities will continue to broaden the range of isotopes for medical applications. For example, the Facility for Rare Isotope Beams (FRIB) at Michigan State University (MSU) will be capable of producing shorter-lived isotopes of key elements for more rapid dose kinetics and new medical applications.

## Future Technologies in Nuclear Medicine

The future impact of nuclear science on medical science is difficult to predict. If history is an indicator, one can expect more significant and exciting contributions. At the least, advances in nuclear medicine will likely remain closely connected with advances in nuclear techniques.

One future direction is personalized medicine, the attempt to identify and treat disorders based on an individual's response to the disease process. This will require more sophisticated nuclear tools. As an example, chemistry systems will be reduced to the size of a postage stamp, thus making patient-specific diagnostic tools and treatment truly individualized. An example of an integrated device, designed for multistep radiosynthesis of PET tracers, is displayed in Figure 3.2.

Other important new directions involve the coupling of advances in genetically engineered antibodies with radionuclides and the use of nuclear imaging to help us understand the underlying causes of disease by extracting functional and anatomical information in the same image.

## MAKING OUR BORDERS AND OUR NATION MORE SECURE

Nuclear science has a long tradition in national security, from the Manhattan Project to today's focus on homeland security. Nuclear devices have determined the outcome of wars and changed the political boundaries of the world. Today, nuclear science plays a critical role in global politics: It protects the borders of the United States, safeguards nuclear material and forestalls the proliferation of nuclear weapons, prevents nuclear terrorism (while at the same time preparing for the "unthinkable"), and ensures that the nation's nuclear weapons stockpile is reliable.

The past decade has seen an expansion in the types of nuclear security problems facing society. For example, considerable effort has been devoted to exploiting new concepts for nuclear forensics and for border protection. At the same time, traditional fields, such as stockpile stewardship and reactor safeguards, have more needs than ever. The contributions from the nuclear physics community to

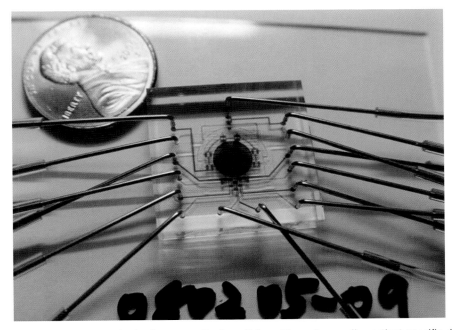

FIGURE 3.2 Future technologies in personalized medicine will require smaller patient-specific diagnostic tools. An example is the chemistry system being designed to produce multiple human doses of FDG, an analog for glucose, on a chip the size of a U.S. penny. In this figure the chip has channels for introducing reagents, channels for opening and closing "pressure valves" by introducing fluids, and channels for venting to allow fluid flow. Flow channels are filled with green dye, control channels with red, and vent channels with yellow. The circle in the center is the reaction chamber. Such devices will reduce time and quantity of reagents and increase efficiency. SOURCE: Courtesy of Arkadij M. Elizarov, Siemens Healthcare. © Copyright Siemens Healthcare 2012. Used with permission.

all of these issues have been both numerous and broad, and a significant number of nuclear physics graduate students express an interest in pursuing careers that address these issues.

## Protecting Our Borders from Proliferation of Nuclear Materials

*Border Detection of Nuclear Contraband*

The priority mission of our nation's Border Patrol is preventing terrorists and terrorists' weapons, including weapons of mass destruction, from entering the United States. Currently there are radiation portal monitors installed at approximately 300 ports of entry. These monitors detect gamma-rays and neutrons emitted from nuclear material. However, one can shield such radiation from detection by placing absorbing material around the nuclear material being smuggled. To deal with such shielding, numerous research groups at universities and national labs are exploring novel detection schemes.

One such scheme scans for high-atomic-number (high-Z) materials hidden in vehicles using cosmic ray muons. As energetic cosmic rays impinge on Earth's atmosphere, they collide with nuclei in the atmosphere to produce copious quantities of muons. Because muons do not interact strongly with the atmosphere, many reach Earth's surface and even penetrate for some distance into shallow mines. Muon radiography takes advantage of this penetrability and is designed to measure the scattering of these muons as they pass through motor vehicles at border inspection stations as a means of detecting hidden nuclear contraband. As sketched in Figure 3.3, the muons are detected both above and below the vehicle. The muons interact with matter in two ways: (1) with atomic electrons, which results in continuous energy loss, and (2) with the atomic nuclei, which results in large angle changes in the muon's path. Each of these interactions provides a radiographic signal that can be used to characterize the material inside a truck. For example, very large angle scattering is a signal that the truck contains high-Z material, such as

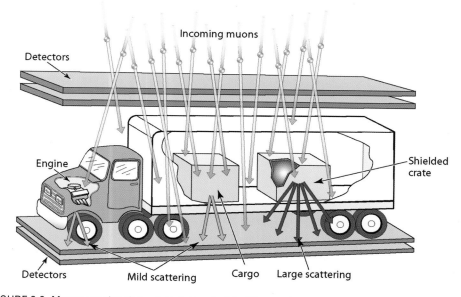

FIGURE 3.3  Muons passing through high-Z materials (like uranium and plutonium) are scattered more than those passing through other materials (such as steel or water). Cosmic ray muons can therefore be used as an active interrogation probe of nuclear materials by detecting muons above and below a truck. SOURCE: Courtesy of C.L. Morris, Los Alamos National Laboratory (LANL).

uranium or plutonium. Muon radiography is proving to be a very efficient border protection tool, and experiments have shown that even high-Z material hidden inside the engine of a vehicle is readily detectable.

## Nuclear Safeguards

The International Atomic Energy Agency's (IAEA's) safeguards system under the Treaty on the Non-Proliferation of Nuclear Weapons, also known as the Nuclear Nonproliferation Treaty (NPT), is aimed at preventing the diversion of civilian nuclear material into military uses. The IAEA safeguards also include schemes for detecting undeclared nuclear activities, such as illicit operations of nuclear reactors. By signing the NPT Treaty, all of the (currently 184) nonnuclear states agree to IAEA safeguard inspections of their nuclear facilities.

One of the very challenging problems for the IAEA is protecting against repeated thefts of small quantities of material over extended time periods. Accountability safeguards largely rely on the detection of gamma-rays and neutrons from nuclear materials, which can be used to deduce inventory anomalies or materials in unauthorized locations. An important component of these schemes is the coupling of advanced radiation detection physics with large nuclear decay databases (and their uncertainties). Scientists at Lawrence Livermore National Laboratory (LLNL) have demonstrated the practicality of gamma-ray nondestructive isotopic measurements using high-purity germanuim (HPGe) gamma-ray detectors. For homogenous materials, one HPGe detector is sufficient to extract isotopic ratio information; for inhomogeneous materials, external transmission sources and multidetector tomography scanning are needed.

## Certifying the Nation's Nuclear Stockpile

### What Happens When Neutrons Interact with Actinides?

To enable certification of the nation's stockpile in the absence of nuclear testing, a number of nuclear physics measurements, coupled with supporting nuclear theory, are being carried out at university and national laboratories. Many of these studies involve neutron-induced cross sections on fissionable actinides (the 14 chemical elements with atomic numbers from 90 to 103, including uranium and plutonium) and other materials that might be found in a nuclear device. Also of importance is the detailed characterization of the energy resulting from fission and fusion.

Current uncertainties on the important fission cross sections for stockpile stewardship are on the order of 2 to 3 percent. In the case of plutonium-239, it would be ideal if this uncertainty were reduced to 1 percent, an improved accuracy also important in developing next-generation reactors. Achieving this level of accuracy requires overcoming uncertainties associated with past fission ionization chamber measurements. Accordingly, a team of LLNL, LANL, and university scientists is developing a fission time projection chamber (TPC), sketched in Figure 3.4, that will be capable of three-dimensional event reconstruction with a high background rejection. Once completed, this will represent a major advance in fission physics. Understanding fission cross sections, especially on actinides other than uranium-235, uranium-238, and plutonium-239 is also important for developing the next generation of nuclear reactors.

In addition to fission, several other neutron-reaction cross sections are needed for stockpile stewardship. Perhaps the most important of these is neutron capture in the tens to hundreds of keV neutron energy region. There is an ongoing neutron capture program involving university and national laboratory scientists and

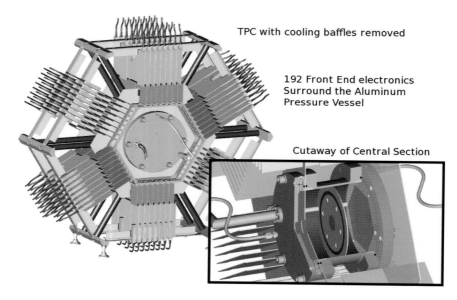

TPC with cooling baffles removed

192 Front End electronics
Surround the Aluminum
Pressure Vessel

Cutaway of Central Section

FIGURE 3.4 TPCs are sensitive instruments in basic research in high-energy and nuclear physics used, for example, in the solenoidal tracker at RHIC (STAR). A new application of a TPC is being developed to enable measurements of neutron-induced fission probabilities of actinides with unprecedented accuracies. The TPC will measure the energy, mass, and direction of fission fragments. Upgrades to the baseline TPC, including additional detectors, would also measure the energy, direction, and multiplicities of fission neutrons and will be able to correlate gamma-radiation with fission events. Such measurements of fission probabilities and properties are important in a wide range of disciplines including nuclear energy, nuclear forensics, national security, and basic nuclear science. SOURCE: Courtesy of M. Heffner, LLNL.

the 4-π $BaF_2$ Detector for Advanced Neutron Capture Experiments (DANCE) at the Los Alamos Neutron Science Center (LANSCE). As an example, some amount of americium-241 is present in all weapons-grade plutonium, and reactions on americium-241 are an important diagnostic for weapon performance. Because the americium-241(n,γ) reaction is important for nuclear forensics, there is a close synergy between the stockpile stewardship and nuclear forensics efforts. DANCE is also used to measure neutron cross sections on unstable targets important for s-process nucleosynthesis.

The extreme conditions in a nuclear explosion result in many of the reactions taking place on unstable nuclei. In the coming decade, access to a much broader range of important unstable isotopes will become possible as FRIB comes online. For many short-lived isotopes, direct measurements will provide information on key reactions of interest. However, it will not be possible to measure all of the relevant reactions, and for the very shortest-lived isotopes theory and simulations will be necessary. FRIB data from related reactions will provide important benchmarking and tests of theory, thus lending confidence to the predictions for the very short-lived unstable isotopes.

### Using Protons to See Where Light Can Never Shine

Over the last decade, proton radiography has become an increasingly important scientific diagnostic tool for weapons science. First demonstrated in 1995, the technique involves using high-energy protons in flash radiography of dynamic experiments, such as implosion tests of mock-ups of nuclear weapons. Protons have advantages over X-rays for certain radiography experiments because protons can penetrate dense materials more efficiently. A key to the success of proton radiography was the realization that magnetic "lenses" can focus the scattered protons to produce exceptionally high-resolution images. The unique feature of proton radiography is its ability to produce high-resolution "movies" of an explosively driven experiment of up to 32 frames, as displayed in Figure 3.5. This allows scientists to probe and quantify dynamic phenomena important in accessing the nation's aging stockpile in the absence of nuclear testing. Today, more than 40 proton radiography experiments are conducted at the LANSCE each year. Other experiments have been carried out at the Alternating Gradient Synchrotron (AGS) accelerator at Brookhaven National Laboratory (BNL).

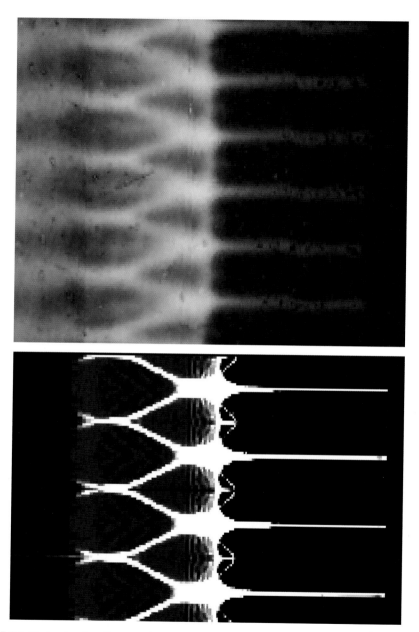

FIGURE 3.5 Understanding the growth of instabilities in shocked material is a major area of research that is being probed with proton radiography. Shown is a comparison of a proton radiograph of spikes and bubbles that are formed by the growth of Rayleigh-Taylor instabilities of a shocked tin surface (*top*) and a hydrodynamic simulation of the experiment (*bottom*). SOURCE: Courtesy of W. Buttler, LANL, and B. Grieves, Atomic Weapons Establishment.

### The Greatest Challenge: Nuclear Devices in the
### Hands of Terrorists or a Rogue Nation

What if the unthinkable happens? A nuclear device is exploded by terrorists or a rogue nation. Or a radiological bomb is detonated in a large U.S. city. Nuclear scientists are using the concepts and tools of nuclear science to assess the risks, monitor for contraband nuclear material, and analyze the devices and materials that could be detonated. The goal is to develop a deterrent for candidate devices for such horrendous actions, as well as to discover what, how, and who should a detonation occur. Nuclear forensics comprises the technical means and set of scientific capabilities that, in the event of an attack, would be used to answer these questions. Nuclear scientists and the tools of nuclear science are keys to addressing the challenges of nuclear forensics, as described in detail in the Nuclear Forensics highlight between Chapters 5 and 6.

## CARBON-EMISSION-FREE ENERGY FOR THE FUTURE

Fossil fuel emissions from power plants foul the air and are central to the discussion of global warming. The emissions contribute to dense brown clouds that hang over cities like Los Angeles and Phoenix, triggering asthma and other respiratory problems. Nuclear energy is an important component of the nation's mission to produce safe, secure, economic, and sustainable energy. Broadly speaking, nuclear energy involves both fission reactors and nuclear fusion. Research in reactor physics spans a broad set of specialties including fuel damage, fuel recycling, safeguards, and waste management. Nuclear physics plays a direct role in addressing each of these. For nuclear fusion, the plasma physics community is exploring a number of methods to achieve the necessary conditions for controlled energy release. In nuclear fusion, plasma conditions approaching those in burning stars are required, and nuclear physics plays a significant role in diagnosing the conditions achieved in the plasma.

### Nuclear Fission Reactors

The majority of the world's nuclear power is generated using reactors based on designs originally developed for naval use. These and other so-called second-generation nuclear reactors are safe and reliable, but they are being superseded by improved designs. Over the past decade, nuclear engineers have been researching advanced reactor designs, and there is a worldwide movement toward to a new generation of reactors. Some of the advanced designs include fast reactors, high-temperature graphite-moderated reactors, thorium-uranium-fueled reactors, pebble bed designs, and mixed oxide fuel (MOX) plutonium reactors. Developing

these designs requires detailed information about the reactions and other physics involved in the processes that are expected to take place. Such measurements
are quite challenging. The fact that many of them are also important to stockpile
stewardship and nuclear forensics greatly enhances our ability to bring together
the teams of scientists needed for these experiments.

### Decay Heat

The energy released during radioactive decay in postfission processes, commonly called "decay heat," accounts for about 8 percent of the energy produced in
the fission process itself. The accurate characterization of decay heat is crucial for
the reactor shutdown process, since it is the main source of heating after neutron-
induced fission is terminated. The decay heat, and in particular the high-energy
part of the radiation, is a key aspect in the proper design of shielding and storage
casks for transporting and storing spent nuclear fuel.[2]

A team of scientists and engineers, led by ORNL, constructed a high-efficiency
modular total absorption spectrometer (MTAS), displayed in Figure 3.6, to measure
the decay heat of fission products. MTAS complements other instruments designed
to directly measure neutron emissions following beta decay of fission fragments,
neutrons that contribute to the neutron budget in a reactor and help to ensure
stable reactor operation. The decay heat and beta-delayed neutron measurements
are also important for understanding r-process nucleosynthesis.

### Reactor Material Damage

Irradiation of both nuclear fuel and structural materials in reactors produces
material defects that limit the safe lifetime of these materials. Numerous irradiation effects can cause material damage, and a number of ongoing collaborations
between nuclear physicists, material scientists, and reactor engineers are examining
and characterizing these effects in detail.

One example is the buildup of helium at grain boundaries and its effect on the
embrittlement of reactor structural materials. The embrittlement of metals such
as nickel, iron, and copper has been demonstrated to be a function of both temperature and helium concentration. Most of the helium is produced by neutron-
induced reactions $(n, \alpha)$, but many of the cross sections for these reactions were not
well known. New cross section measurements have resulted in significant changes
in estimates of the probable safe lifetime of structural reactor materials.

In nuclear fuels, a major cause of damage is the buildup of bubbles of noble

---

[2] The text of this paragraph is adapted from K.P. Rykaczewski, 2010, Viewpoint: Conquering nuclear
pandomenium, *Physics* 3:94.

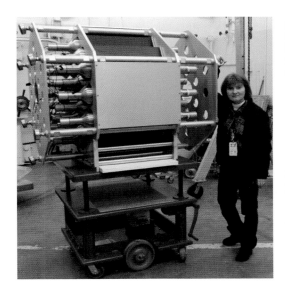

FIGURE 3.6 In a nuclear reactor, a sizeable fraction of the energy when the reactor is on and after it is shut down comes from the radioactive decay of fission products, also known as decay heat. A state-of-the-art device is being commissioned at ORNL to measure the decay heat of fission fragments with high accuracy. The MTAS consists of 19 NaI(Tl) modules with 48 photomultipliers (left panel). Its total volume will be about seven times that of the largest existing total absorption spectrometer. The right panel shows the first MTAS crystal manufactured at Saint Gobain Crystals, in Hiram, Ohio. In the coming decade, FRIB will produce a greatly expanded set of fission fragments and enable precision measurements of their detailed decay modes. SOURCE: Courtesy of K. Rykaczewski and M. Wolinska-Cichocka, ORNL.

gases and their migration through the fuel. Gas bubbles can cause changes in internal gas pressure, thermal conductivity, temperature gradients, and material stress and strain, thus inducing damage or even failure in fuel and cladding materials over time. Understanding the formation and properties of these bubbles and how to detect the gases if released is the focus of a joint collaboration between materials and nuclear scientists.

*Fuel Performance and Next-Generation Reactors*

One advanced concept is the fast reactor, wherein the neutron flux is considerably higher in energy than in standard thermal reactors. The dominant neutron energies for a fast reactor are 0.1-0.6 MeV. Fission cross sections are considerably less well known at fast reactor energies than at thermal energies. And the situation is most serious for the transuranic fuels. New programs are under way to measure the fission cross sections, where experimentally feasible, on less abundant isotopes of plutonium and uranium, as well as the minor actinides such as isotopes of americium, curium, and neptunium. In addition to the fission cross sections, accurate knowledge of the neutron capture cross sections on the minor actinides is important. Many of these actinides are radioactive, restricting measurements to small targets. International collaborations are addressing these problems using the DANCE detector at LANSCE, which is designed to study neutron capture reactions on small quantities, about 1 mg, of radioactive and rare stable nuclei. Others are using the TPC displayed in Figure 3.4 to determine fission cross sections with unprecedented accuracy.

One major attractive characteristic of fast reactors is their enhanced ability to burn up highly toxic transuranic fuel produced as waste from light water reactors. At these higher neutron energies, there are a number of nuclear properties of reactor fuels that need to be determined to considerably higher accuracy than is presently possible, including reactions of neutrons with unstable fission products. In the future, FRIB will extend capabilities by allowing studies of a considerably larger class of unstable isotopes. For several key nuclides that have longer half-lives, FRIB will provide separated samples that can be used to measure neutron capture probabilities at neutron beam facilities. For isotopes with shorter lifetimes, indirect reaction measurements at FRIB will provide information to help constrain theoretical models for neutron-capture probabilities, using techniques that will also advance basic nuclear science and nuclear astrophysics.

## Nuclear Fusion Energy

When two light nuclei interact, they can fuse to form a heavier nucleus, accompanied by the release of a large amount of energy. The conditions found in stellar environments are ideal for sustained fusion chains, and our sun is a natural fusion reactor. However, achieving these hot, dense conditions in the laboratory is very challenging, and to date the only successful terrestrial events have been thermonuclear explosions. Currently there are two main research approaches to fusion: magnetically driven fusion and laser-driven fusion. The National Ignition Facility (NIF) at LLNL is a laser-driven inertial confinement fusion (ICF) facility.

### High-Energy-Density Physics

Probing physics at high energy densities is central to several subfields of nuclear physics, including the study of nucleosynthesis, the quark-gluon plasma, and neutron stars. The NIF provides a unique regime in the temperature-density (T-$\rho$) plane of high-energy-density physics (see Figure 3-7). NIF is designed to compress capsules containing a mixture of deuterium (d) and tritium (t) to temperatures and

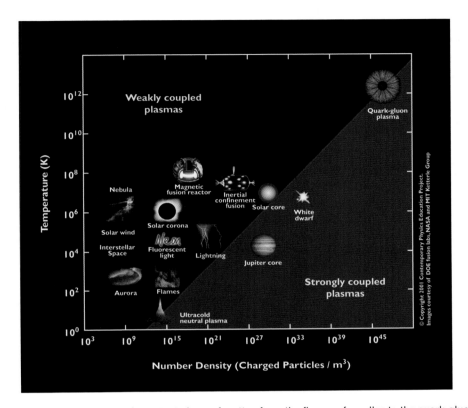

FIGURE 3.7 Plasmas are an important phase of matter, from the flames of candles to the quark-gluon plasma generated for a fraction of a second in a relativistic heavy ion collision. The temperature in kelvins as a function of the number of charged particles per cubic meter for a wide range of physical systems is displayed. The National Ignition Facility produces plasmas via inertial confinement fusion that are comparable to the interior of the sun. SOURCE: Courtesy of the Contemporary Physics Education Project.

densities high enough to ignite thermonuclear reactions. Laser pulses, directed into a hohlraum cylinder containing the target capsule, create an X-ray bath sufficient to compress the capsule through ablation of an outer layer of material. Achieving the conditions needed for ignition is challenging but made more tractable with the use of advanced diagnostics, many of which are based on nuclear physics.

*Using Neutrons to See Fusion*

The main fusion reaction at NIF is the d + t $\Rightarrow$ n + $\alpha$ reaction, which releases 17.6 MeV of energy per reaction in the form of a 14-MeV neutron (n) and a 3.6-MeV alpha particle. If successfully ignited, an NIF capsule will burn about $10^{18}$ d + t

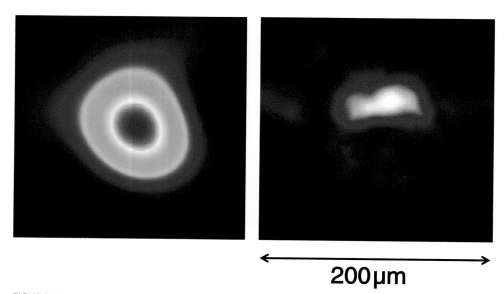

**200 µm**

FIGURE 3.8 The NIF at LLNL is striving to use high-powered lasers to fuse deuterium and tritium, recreating in the laboratory the source of energy in our sun. Neutrons are one of the main products of the fusion reactions. Nuclear physicists are developing tools to diagnose the conditions in the NIF d-t capsules. On the left is a simulation of an expected neutron image; on the right is a reconstruction of an actual neutron image of a capsule taken at NIF. The image determines the size of the hotspot and the asymmetry of the implosion. SOURCE: NIF, LLNL.

reactions, with a corresponding release of over 2.5 MJ of fusion energy. One of the important diagnostics for understanding capsule behavior is neutron imaging. The system being built for NIF will image both the primary 14-MeV neutrons from the d + t reaction and the lower energy 6- to 13-MeV neutrons resulting from these primary neutrons losing energy in the material, or downscattering. The image of the primary 14-MeV neutrons determines the size of the burning fuel region (the hotspot). The lower-energy, downscattered neutrons provide information on the average density as a function of radial distance from the center of the fuel and on how symmetric or asymmetric an implosion was achieved. The imaging system, a set of pinhole apertures placed close (within 20 cm) to the capsule and an imaging camera placed far (28 m) from the capsule, will be capable of imaging neutrons from capsules that burned as few as $10^{14}$ d + t reactions, as shown in Figure 3.8.

## INNOVATIONS IN TECHNOLOGIES AND
## APPLICATIONS OF NUCLEAR SCIENCE

Nuclear physics is fundamentally cross-disciplinary in nature, providing experimental and theoretical tools and concepts for countless other sciences and contributing to the quality of life across a wide spectrum of social and economic needs. The applications and manifestations are so entrenched in our daily lives as to be ubiquitous, from simple everyday household items to technologies that provide significant portions of the foundation of medical procedures. Nuclear science has and will continue to play a substantial role in developing solutions for energy, climate, and environmental challenges. Further, the primary tools of modern nuclear science—accelerators and computers—have spawned many applications and economic benefits, some of which are discussed here.

## Addressing Challenges in Medicine, Industry, and Basic Science with Accelerators

Beams of high-energy particles, produced by accelerators, are essential for both fundamental and applied research and for technical and industrial fields. Accelerators have become prevalent in our lives, and there are now over 30,000 accelerators worldwide. Of these, the largest number (about 44 percent) are used for radiotherapy, while 41 perecnt are used for ion implantation, 9 percent for industrial research, and about 4 percent for biomedical research. The remaining 1 to 2 percent of accelerators are very high-energy accelerators used in nuclear and particle physics to probe the fundamental nature of the matter making up our universe.

All accelerators can be described as devices that use electric fields to accelerate charged particles (such as electrons or ions) to high energies, in well-defined beams. Since the discovery of the X-ray in 1895 by Roentgen, many famous nuclear physicists have made seminal contributions to new accelerator technologies, including John D. Cockcroft, Ernest Walton, Earnest O. Lawrence, and Robert Van de Graaff. Today accelerator technologies range from the Large Hadron Collider (LHC) capable of producing TeV particles to the lowest energy accelerators used by industry.

### Accelerators and Medicine

Accelerators form the basis for many diagnostic systems, from chest X-ray machines to whole-body X-ray scanners capable of creating a three-dimensional image of the living body. Accelerators such as cyclotrons enable protons and other light nuclei to be used to produce radioactive nuclei that are used in diagnostic medicine. Radioisotopes such as thallium-201 are used to diagnose heart disease. The production of the unstable isotopes of the elements of life, such as oxygen-15, carbon-11, nitrogen-13, and the pseudo-hydrogen fluorine-18, has led to the field of PET. These positron-emitting radionuclides are attached to biologically active molecules. When the tagged molecules are injected, the annihilation radiation can be imaged and the functional capacity of the patient can be determined, as discussed in the PET highlight, located between Chapters 2 and 3. Today PET scanners are combined with computed tomography (CT) scanners so that in one setting, the structural (CT) and functional (PET) capacity of the patient can be determined. CT and PET scanners have revolutionized nuclear medicine.

Intense X-rays are now one of the primary modes of treating cancer. Accelerators throughout the world generate beams of electrons that are directed to targets that create X-rays, which are then directed at the tumors to destroy them. The modern therapy machine has become extremely sophisticated in that the electron beam can be modulated to increase and decrease the flux to alter the dose of X-rays and thereby spare healthy tissue while maximizing the dose to the tumor. While the standard of care for cancer treatment includes X-ray therapy, there is a growing use of high-energy protons to ablate the tumors. The idea is to deposit as much energy as possible in the tumor cells while sparing the surrounding tissues.

In the United States, partnerships between industry and nuclear science laboratories have led to new accelerator developments for medical applications. For example, the National Superconducting Cyclotron Laboratory (NSCL) at Michigan State University has pioneered the application of superconducting accelerator technology in medicine. This work has resulted in the miniaturization of the cyclotron so that it will fit on a gantry and rotate around the subject, simplifying beam delivery and allowing for tighter control of radiation dose delivery. NSCL has also designed and constructed a gantry-mounted, superconducting K100 medical cyclotron, funded by Harper Hospital in Detroit, for neutron therapy. The NSCL's conceptual design for a superconducting cyclotron for proton therapy has been adopted and further refined by Varian Medical Systems/ACCEL Corporation, with technical advice from NSCL faculty and staff.

The success of proton therapy has stimulated interest in using heavier hadrons, such as carbon ions, with the potential of depositing more energy to a small area. Several synchrotrons delivering carbon-12 for therapy have been installed in Europe and Japan. At Brookhaven National Laboratory (BNL), home of the Relativistic Heavy Ion Collider (RHIC), next-generation accelerators for precise, safe cancer radiotherapies are being developed.

### Accelerators in Industry and for Energy

There is a vast enterprise of techniques that use accelerators in a wide range of industries to polymerize plastics, to sterilize food and medical equipment, to weld materials using an electron beam, to implant ions into materials, to etch circuits on electronic devices, to examine the boreholes of oil wells, and to search for dangerous goods. There are approximately 8,500 such devices worldwide.

Electron beams dominate the industrial uses, with the curing of wire-cable tubing and of ink accounting for more than 60 percent of the market. Other electron beam uses include shrinking films, cross bonding of fibers in tires, and irradiation of food. Here, electron beams replace traditional thermal heating approaches because of the gain in efficiency that comes from the more uniform distribution of energy.

A number of major accelerator developments related to nuclear energy are being pursued, including plasma heating for fusion reactors, inertial fusion reactors, nuclear waste transmutation, electronuclear breeding, and accelerator-driven subcritical reactors.

### Basic and Applied Science

The breadth of scientific disciplines that make use of accelerators to perform their studies is considerable. Cutting-edge materials research makes use of synchrotron radiation having a wide range of wavelengths. Muon beams and neutrons produced from spallation sources probe the properties of materials such as the high-temperature superconductors. Mass spectroscopy is a standard analytical technique for chemists. As discussed at the end of this chapter, high-resolution mass spectrometry is used in archaeology and geology for dating artifacts by determining the ratio of stable to long-lived isotopes.

### Free-Electron Lasers

A free-electron laser (FEL) is a powerful source of coherent electromagnetic radiation that is produced by a relativistic electron beam propagating through a periodic magnetic field (see Figure 3.9). FELs are capable of producing intense radiation over a wide range of the electromagnetic wave spectrum, from microwave to hard X-ray, with average beam powers up to tens of kilowatts and peak powers up to tens of gigawatts. FELs are used for research in many fields, including materials science, surface and solid-state physics, chemical, biological and medical sciences, and nuclear physics. While the principle of operation of all FELs is the same, each device is optimized for its main application. FELs that are used in applications that require high average power are typically operated in the infrared (IR) region and are driven by a high-repetition-rate linear accelerator with an optical resonator. Nuclear physics accelerator facilities are leading new developments in FEL technologies.

New investigations in condensed matter studies at accelerator labs in the United States and Germany have already identified previously unknown interstellar molecular emission lines, developed new processes for production of boron nitride nanotubes, and produced nonthermal pulsed laser deposition of complex organics on arbitrary substrates. Superconducting radiofrequency technology developed at the Continuous Electron Beam Accelerator Facility (CEBAF) nuclear physics accelerator is now being commercialized for future implementation in weapons

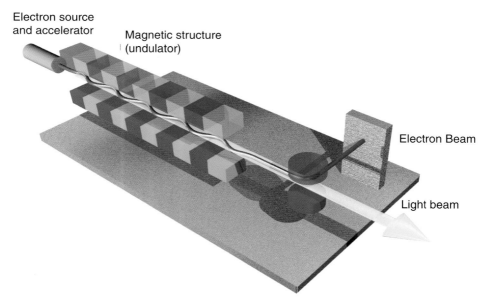

FIGURE 3.9 FELs are a powerful source of coherent electromagnetic radiation that is produced by a relativistic electron beam propagating through a magnetic field. They are used in numerous basic and applied science applications, including probing materials, biological systems, and nuclei. Shown is a schematic diagram of the basic layout of an FEL. The electron beam is transported through the periodically varying magnet field of an undulator magnet. Microbunching inside the electron beam at a spacing equal to that of the light's wavelength enables electrons to radiate coherently in order to establish lasing. An FEL can be operated with either an optical resonator or in a single-pass configuration with a long undulator section. SOURCE: Image courtesy of Deutsches Elektronen-Synchrotron (DESY) in Hamburg, Germany. Copyright: DESY 2006.

systems for the U.S. Navy. And FEL technologies and applications are strongly coupled to nuclear physics research, including the technologies needed for a future electron-ion collider.

## Information and Computer Technologies

Both nuclear physics experiments and theory have been enabled by and, in turn, have spawned, advances in computer science and technology. For experimentalists, the enormous quantity of data that characterize modern nuclear physics experiments has required that systems be devised to handle and make such data meaningful. RHIC experiments now routinely collect petabyte scale data sets each year, at rates of 1 GB per second. Analysis of such data sets drives technology development for the sustained use of data grids. For example, the computing groups for the STAR collaboration at RHIC have developed a data movement service to achieve sustained and robust automated data transfers of 5 TB a week, with peak data transfer rates reaching 30 MB per second. This allows next-day access to fresh data from the experiments for analysis.

Analogous progress has come out of the need for massive and reliable computational approaches to address some of the fundamental problems in nuclear theory. Lattice quantum chromodynamics (QCD) calculations of the structure and properties of protons and hot quark-gluon plasmas that begin with fundamental quark and gluon building blocks are among the most demanding numerical computations in nuclear physics. Advancing this basic science drives innovation in computer architectures. In a lattice QCD calculation, space and time are rendered as a grid of points, and the quarks and gluons at one point interact only directly with those at other nearby points. This localization of the particles and their interactions makes these numerical computations particularly well suited for massively parallel supercomputers, with communications between processors having a simple pattern that enables the efficient use of a very large number of processors.

This characteristic of lattice QCD calculations drove some physicists to design special-purpose supercomputers that attracted attention in the broader computer hardware arena by achieving lower price-to-performance ratios than contemporary commercial supercomputers. A particularly successful group designing special-purpose lattice QCD supercomputers was based at Columbia University, working in partnership with IBM, which manufactured the computer chips. Originally, the group built a machine based on a low-power, simple, digital signal-processing chip (similar to those in cell phones) and a special-purpose serial communication network. This partnership laid the foundation for a new machine called the QCDOC (QCD on a chip), displayed in Figure 3.10, in which the whole processing unit, including a newer more powerful microprocessor, the communication network, and memory, was integrated on one chip.

Recently, the LHC, which enables particle and heavy-ion nuclear physics research at the energy frontier, has reached unprecedented volumes of data and requirements for data transfer rates and data processing power. This has led to the development of technology that allows extraordinary data transfer rates at large distances. At Super Computing 2011, the International Conference for High Performance Computing, Networking, Storage and Analysis, held in Seattle, Washington, in November 2011, a new world record of bidirectional data transfer rate was achieved: 23 GB per second between the University of Victoria Computing Centre located in Victoria, British Columbia, and the Washington State Convention Center in Seattle. Such technology eventually will influence the Internet infrastructure used in our everyday life.

Lattice QCD machines, QCDOC in particular, became the paradigm for a new generation of world-leading massively parallel supercomputers that are currently

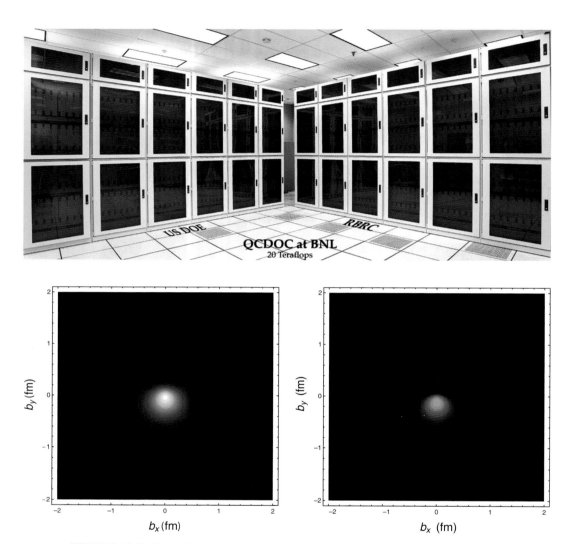

FIGURE 3.10 Nuclear science computing needs have led the community to develop new and innovative communication networks, data transport and manipulation systems, and computer architectures. An example is the QCDOC supercomputer at BNL (shown in the upper figure), a joint venture between RIKEN in Japan and the U.S. Department of Energy, in partnership with IBM. Examples of calculations now possible with the most powerful computers are given in the lower figures. Displayed are lattice QCD calculations of the transverse charge distributions of a proton (lower left) and a neutron (lower right), polarized in the x-direction, as a function of the radial distance from the center of the nucleon computed. These transverse charge densities are shown in a reference frame in which the observer is riding along with the photon (the Breit frame). In both cases, the charge distribution has an electric dipole component in the y-direction. This effect is entirely due to the interplay of special relativity and the internal structure of the nucleon. SOURCE: *(top)* Courtesy of Brookhaven National Laboratory; *(bottom)* Courtesy of Huey-Wen Lin and Saul D. Cohen, University of Washington.

being used in a vast array of applications having impacts in science and on the broader economy. In particular, IBM built the successful commercial Blue Gene line of computers, which engaged several former Columbia students and postdoctoral scholars. In addition to lattice QCD calculations, these supercomputers have been just as successful in simulating exploding stars or nuclear reactors, both of which require enormous computing power. Climate science researchers at BNL are using a Blue Gene named New York Blue to make significant progress in understanding today's climate and to better predict climate evolution. Genomic sequencing, protein folding, materials science, and brain simulations are also prominent on the list of successful Blue Gene applications. Special-purpose supercomputers for lattice QCD have also been designed in Europe (the Array Processor Experiment) and Japan (the CPPACS and PACS-CS projects in Tsukuba).

### Cosmic Rays, Electronic Devices, and Nuclear Accelerators

Cosmic rays are continuously bombarding Earth: more during active solar periods, more at the poles, and less at the equator. When cosmic rays, or radiation from their secondary products, interact with an electronic device, the function of that device can be compromised. The resulting errors in the functionality of an electronic device, such as the one displayed in Figure 3.11, can have very serious consequences for technologies used by such disparate industries as aerospace and autos.

A single event upset (SEU) refers to a change in the state of the logic or support circuitry of an electronic device caused by radiation striking a sensitive location or node in the device. SEUs can range from temporary nondestructive soft errors to hard error damage in devices. The detailed physics determining the rate at which SEUs occur is both complicated and device dependent. Circuit manufacturers try to design around the risks posed by cosmic ray interactions by introducing redundancy or other protective measures to compensate for the radiation-induced errors. To do so requires detailed knowledge of the expected rates and types of SEUs that can occur. Thus, experimental testing of semiconductor device response to radiation requires beams of particles that provide realistic analogs of cosmic rays and their secondary products. The main particles responsible for SEUs are neutrons, protons, and alpha particles, as well as heavy ions. Thus, the beams needed for this large experimental program require a range of nuclear accelerator facilities to test for device vulnerabilities and to characterize the radiation-induced failure modes of the electronic chips. For this, nuclear physics accelerator facilities are a unique resource, and agencies and companies from all over the world purchase beam time at accelerator facilities to test for device vulnerabilities and to characterize the radiation-induced failure modes of the electronic chips. In the United States

FIGURE 3.11 Nuclear physics laboratories across the world are working in collaboration with the aerospace and semiconductor industries to assess the impact of cosmic rays on electronic devices such as computer chips. Ongoing research programs are involved in testing the effects of heavy ions and neutrons on microelectronic devices, such as this one being studied at Texas A&M University. SOURCE: Zig Mantell and Texas A&M University.

alone, each year national and university nuclear physics laboratories provide almost 10,000 hours of accelerator time for this important service.

## Helping to Understand Climate Effects One Nucleus at a Time

Applications of nuclear techniques are used to advance other scientific disciplines, including climate science, cosmochemistry, geochronology, paleoclimate, paleo-oceanography, and geomorphology. Since 1949, when Willard Libby first demonstrated carbon dating, the field of trace analyses of long-lived cosmogenic isotopes has steadily grown. Because they are chemically inert, noble gases play a particularly important role as tracers in environmental studies. Owing to their inertness, the geochemical and geophysical behavior of these gases and their distribution on Earth is simpler to understand than that of reactive elements. In addition, their inertness facilitates recovery of minute quantities from very large volumes of other material. Precision tools and techniques developed for basic nuclear physics continue to be applied to answer open questions in climatology, geology, and oceanography.

*Probing Ancient Aquifers in Egypt*

A challenging problem in earth science is the determination of the residence times and flow velocities of groundwater circulating deeply through Earth's crust. Krypton-81, which is produced by cosmic-ray-induced spallation in the atmosphere, has been identified as an ideal chronometer for determining fluid residence times on the $10^5$-$10^6$ year timescale. However, since krypton-81 is such a rare isotope it has been extremely difficult to measure its abundance.

A new method, atom trap trace analysis (ATTA), was developed at Argonne National Laboratory to analyze krypton-81 in environmental samples. With a half-life of 230,000 years and an atmospheric isotopic abundance of one part per trillion, krypton-81 can provide unique information on terrestrial issues involving million-year timescales. Individual krypton-81 atoms can be selectively captured and detected with a laser-based atom trap. Joining low-level counting and accelerator mass spectrometry (AMS), two methods previously developed by nuclear physicists, ATTA is the newest method to detect tracers with an isotopic abundance at parts per trillion.

Using ATTA, krypton-81 atoms in environmental samples can now be counted and the isotopic abundance of krypton-81 measured. In the first application of ATTA to a groundwater study, a team of geologists and physicists from the United States, Switzerland, and Egypt sampled krypton from the Nubian Aquifer groundwater (displayed in Figure 3.12), which is of unknown age. Following extraction of krypton from thousands of liters of water at six deep wells, the krypton-81/Kr ratios measured by ATTA indicated groundwater ages ranging from 200,000 to 1,000,000 years. These results characterized the age and hydrologic behavior of this huge aquifer, with important implications for climate history and water resource management in the region. The success of this project suggests that widespread application of krypton-81 in earth sciences is now feasible.[3]

*Tracing Ocean Circulation*

It is becoming more apparent that the oceans are a major regulator for our world's climate. One of these "motors" is the Atlantic conveyor belt system, whereby

---

[3] Portions of the discussion in this section are adapted from the Argonne National Laboratory, 2003, *Physics Division Annual Report*, Chapter IV.

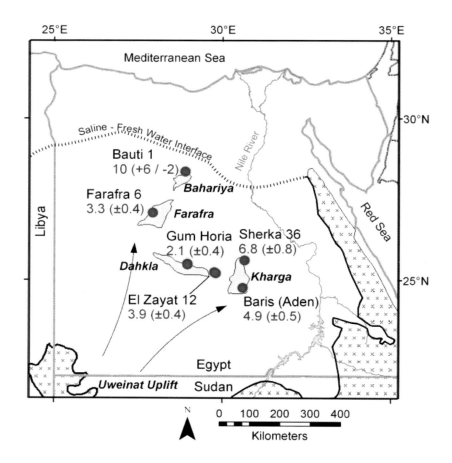

FIGURE 3.12 Understanding the flow of groundwater that circulates through Earth's crust is an open question in geology. In a collaboration of nuclear scientists and geoscientists, the precision technique of atom-trap analysis was used to measure the radioactive isotope krypton-81 in deep wells of the Nubian Aquifer in Egypt. The map shows sample locations and their krypton-81 ages (in 100,000 years) in relation to oasis areas (shaded green). Groundwater flow in the Nubian Aquifer is toward the northeast. SOURCE: Adapted from N.C. Sturchio et al., 2004, One million year old groundwater in the Sahara revealed by krypton-81 and chlorine-36, *Geophysical Research Letters* 31. Copyright 2004 American Geophysical Union. Reproduced/modified by permission of American Geophysical Union.

warm water is chilled in the far North Atlantic, sinks to greater depths, and flows down the Atlantic across the Indian Ocean into the Pacific, where it heats up, rises to the surface, and flows back to the North Atlantic, as displayed in Figure 3.13. The whole cycle takes about 1,000 years. It is becoming increasingly clear that the amount of heat transported from the tropics to the polar regions by the oceans

FIGURE 3.13 Thermohaline circulation, commonly referred to as the ocean "conveyor belt," is made up of ocean currents that transport heat from the tropics to the polar regions. AMS of the radioactive isotope argon-39 will be used to explore this conveyor belt and its impact on climate. SOURCE: National Oceanic and Atmospheric Administration.

is comparable to the amount transported by the atmosphere. Therefore, it is very important to understand this system. With a half-life of 269 years, argon-39 is particularly well suited to study questions related to ocean circulation. However, its extremely low concentration (argon-39/Ar = $8.1 \times 10^{-16}$), coupled to its long half-life, makes it impossible to measure the argon-39 decay in any sample of reasonable size.[4]

AMS using the ATLAS heavy ion accelerator at Argonne National Laboratory has been successful in separating argon-39 from its ubiquitous potassium-39 isobaric background, the latter being 6-7 orders of magnitude more intense. Measurement of isotopic ratios as small as argon-39/Ar = $4 \times 10^{-17}$ have been achieved. This program is now poised to measure argon-39 concentrations in ocean water samples in order to explore the oceanic "conveyor belt."

---

[4]Portions of this paragraph have been adapted from M. Gaelens, M. Loiselet, G. Ryckewaert, et al., 2004, Oceans circulation and electron cyclotron resonance sources: Measurement of the AR-39 isotopic ratio in seawater, *Review of Scientific Instruments* 75: 1916.

## Highlight: Future Leaders in Nuclear Science and Its Applications: Stewardship Science Graduate Fellows

To address compelling concerns in national and homeland security requires highly trained, talented individuals in a variety of disciplines, including nuclear science. The Department of Energy's National Nuclear Security Administration (NNSA) has identified the need to develop a highly talented workforce of U.S. citizens to meet the long-term requirements of national security and the stewardship mission of NNSA. To meet this need, the NNSA established the Stewardship Science Graduate Fellowship program in the areas of high-energy-density physics, materials under extreme conditions, and low-energy nuclear science. In addition to receiving stipends and tuition remission, all fellows are required to spend at least 3 months in residence at one of the NNSA laboratories for a practicum. Of the 23 fellows in the first 5 years of the program, 8 are doing research in nuclear science. The stories of four of these fellows are told here.

To find a robust energy source with minimal emission of greenhouse gases will require a renewed commitment to nuclear energy and developing the next generation of nuclear reactors. Preparing for this next generation of nuclear reactors requires an understanding of the probability that actinides other than uranium-235 or plutonium-239 will fission. As a part of his Ph.D. dissertation work, Paul Ellison, a fellow from the University of California at Berkeley, is developing techniques to measure the fission probability of the rare isotope americium-240, which has a half-life of only 51 hours (Figure FEL 1). He will be creating this isotope with a nuclear

FIGURE FEL 1 Graduate student Paul Ellison with the Berkeley Gas-Filled Separator, used for studying the chemistry and physics of the heaviest elements. SOURCE: Image courtesy of PT Lake.

reaction in which plutonium-242 is bombarded with protons at the energy that maximizes the emission of three neutrons to make americium-240. Once he has mastered the techniques to produce the tiny amounts (<100 ng) of americium-240 required to measure fission probabilities, experiments will be performed at the LANL Neutron Science Center, where he has already spent several months as part of his practicum. Paul also performs fundamental low-energy nuclear physics research on the heaviest elements, such as the yet unnamed element-114.

The nucleus is a complex quantum system made of neutrons and protons. One goal of nuclear structure research is to establish a unified framework for understanding the properties of atomic nuclei and, potentially, for extrapolating to the limits of nuclear existence. Fellow Angelo Signoracci from MSU is working on such a unified framework (Figure FEL 2). He will be developing a hybrid method that builds on decades of experience with shell model calculations, which provide detailed predictions of nuclear structure with many input parameters, and the energy density functional approach, which could predict the entire nuclear landscape with one parameterization. By spending his practicum at LLNL, he was able to interact with the laboratory's scientists, who are leading efforts that exploit its petascale computing facilities.

FIGURE FEL 2 Graduate student Angelo Signoracci develops computer models of the structure of atomic nuclei. SOURCE: Image courtesy of K. Kingery, Communications manager at MSU's National Superconducting Cyclotron Laboratory.

*continued*

The ability to detect low levels of radiation is important for many applications and critical for detecting weakly interacting particles such as neutrinos and dark matter. Nicole Fields, a fellow from the University of Chicago, is developing and characterizing P-type point contact (PPC) germanium detectors and associated electronics (Figure FEL 3). These detectors will be the primary design component of the Majorana experiment that will search for neutrinoless double-beta decay and light-mass dark matter candidates. She also works closely with industry in improving these detectors. Upon completing 2 years of graduate studies, Nicole looks forward to doing her practicum.

The detection of neutrons and understanding reactions induced by neutrons and protons are important for basic nuclear structure and astrophysics, as well as for applications in home-

FIGURE FEL 3 Graduate student Nicole Fields is developing new systems to detect weakly interacting particles. SOURCE: Image courtesy of Lloyd DeGrane, University of Chicago.

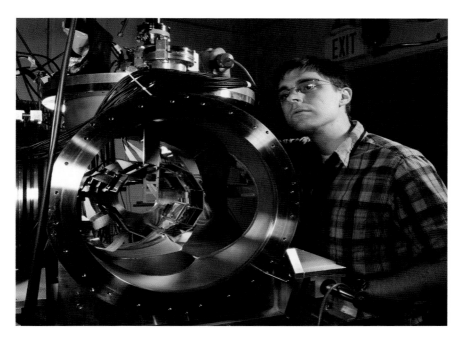

FIGURE FEL 4  Patrick O'Malley uses the ORNL accelerator and the ORRUBA array of position-sensitive silicon strip detectors to study nuclear reactions that help us understand how the elements are synthesized in stars. SOURCE: Patrick O'Malley.

land security. The research of fellow Patrick O'Malley from Rutgers University is focused on nuclear reaction studies and in particular on the question of how the stable isotope fluorine-19 is synthesized in stars when nitrogen-15 interacts with alpha particles (Figure FEL 4). Patrick measured the reaction in which nitrogen-15 interacts with deuterium, emitting a proton and thereby transferring a neutron to form nitrogen-16. He hopes to understand the destruction of nitrogen-15 from interactions with neutrons in stars. Much of his research involves the charged-particle detectors of the ORNL-Rutgers University Barrel Array (ORRUBA). He is also part of the team developing the Versatile Array of Neutron Detectors at Low Energy. Patrick spent a summer at LLNL helping to develop the neutron time projection chamber, which can be applied to locate sources of special nuclear materials that emit neutrons.

# 4

# Global Nuclear Science

The G-20 nations—Argentina, Australia, Brazil, Canada, China, France, Germany, India, Indonesia, Italy, Japan, Mexico, Russia, Saudi Arabia, South Africa, South Korea, Turkey, the United Kingdom, and the United States, plus the European Union—include countries with advanced economies and countries with emerging economies that are working together toward worldwide financial stability and the achievement of sustainable growth and development. As is generally recognized in the G-20 member nations, a crucial component for creating an innovative economic environment is a commitment to invest in research and education. This commitment is coupled with appropriate national policies in science and technology to attract the brightest minds and allow them to create new technologies to benefit society. Through the development of research facilities, the United States has enjoyed global leadership and dominance in many of the basic sciences. With increasing demands on resources from large science projects, a strategy has begun to emerge where members of the G-20 collaborate on the construction and operation of some of the largest projects. Examples of this include the ITER fusion reactor project and the Large Hadron Collider (LHC) at the European Organization for Nuclear Research (CERN).

Today, most of the G-20 member nations support active programs in nuclear science. Other countries, especially those in the European Union, also support both facilities and programs in nuclear science. Below is a brief description of some of the major nuclear science programs and research facilities found in these countries. Table 4.1 provides information about the major facilities in the United States and Table 4.2 provides similar information about facilities abroad. The report is not

Table 4.1 Domestic Nuclear Physics Facilities

| Facility | Beam Characteristics | | Research Areas | Number of Users per Year | Future Upgrades |
|---|---|---|---|---|---|
| | Species | Energy | | | |
| ANL ATLAS Argonne, Ill. | Protons, heavy ions ($1 \le A \le 238$), some rare isotope beams | <18 A MeV | Study of atomic nuclei near and far of stability and at high spin, nuclear astrophysics, and fundamental symmetries with stable and radioactive beams. Accelerator physics. | 411 | CARIBU facility for stopped and reaccelerated fission products. |
| JLAB CEBAF Newport News, Va. | Electrons | 1-6 GeV | Probe the nucleus to understand quark matter. | 1,206 | Energy range increase to 12 GeV for better quark matter research. |
| | Free-electron laser | 10 kW (IR) | Superconducting radiofrequency (RF) accelerator development. | | FEL upgrade to 1 kW in the UV range. |
| MSU NSCL East Lansing, Mich. | Protons, heavy ions ($1 \le A \le 238$), wide range of rare isotope beams | <200 A MeV | Study of atomic nuclei very far from stability, nuclear matter, nuclear astrophysics, and fundamental symmetries with radioactive beams. Accelerator physics. | 718 | ReA3 and ReA12 facilities for gas stopping and reacceleration of radioactive beams to 3 A MeV and 12 A MeV, respectively. Recoil separators. |
| BNL RHIC Upton, N.Y. | Heavy ion collider ($d \le A \le Au$) | (maxima) 100 + 100 A GeV (equivalent to fixed-target collisions at 21,000 A GeV) | Create, explore, and understand matter at extreme temperatures and energy densities governed by quantum chromodynamics (QCD). | 1,100 | Increasing RHIC's luminosity. Detector upgrades. |
| | Proton collider | 250 + 250 GeV | Analyze behavior of gluons, quarks and antiquarks in protons. | | |

SOURCE: Information contained herein is based on the IUPAP Worldwide Overview of Research Facilities in Nuclear Physics, Booklet 41, available online at http://www.triumf.info/hosted/iupap/icnp/Report41-final-12-07-11.pdf. Except for MSU NSCL, the number of users per year is from the Summary Table, page xxix. The number of users per year for MSU NSCL is taken from the body of Booklet 41, given that the figure in the Summary Table is significantly lower than that listed in the body of the booklet.

TABLE 4.2 International Nuclear Physics Research Facilities

| Facility | Beam Characteristics | | Research Areas | Average Number of Users per Year | Future Upgrades |
| | Species | Energy | | | |
| --- | --- | --- | --- | --- | --- |
| CERN Franco-Swiss border | Pb-Pb collisions, <br><br> p-p collisions | 2.76 A TeV <br><br> 7 TeV | Studies of quark-gluon plasma. <br><br> Determination of the Higgs boson and supersymmetry for verification of Standard Model. <br><br> Investigations into dark matter. | 10,000 (for whole CERN facility) | SuperLHC luminosity upgrade in 2018. <br><br> 7 TeV/beam upgrade scheduled for 2014. |
| CERN, ISOLDE Franco-Swiss border | Radioactive ions | 3 A MeV | Production, study, and acceleration of radioactive nuclei (ISOLDE). | 10,000 (for whole CERN facility) | HIE-ISOLDE, a beam energy, intensity, and flexibility upgrade, including a new linac <br><br> SCREX-ISOLDE, a superconducting upgrade to the REX-ISOLDE experiment |
| GANIL Caen, France | Heavy ions ($12 \leq A \leq 130$) <br><br> Exotic beams (ISOL) | 95 A MeV <br><br><br> 25 A MeV | Nuclear structure, including reactions and properties. <br><br> Studies of exotic nuclei. | 370 | SPIRAL2 radioactive beam facility is under construction. |
| GSI Helmholtz Centre for Heavy Ion Research Darmstadt, Germany | Protons <br><br> Heavy ions/ radioactive ions <br><br> Pions | 4.7 GeV <br><br> 1.4-12 A GeV <br><br><br> 0.5-2.5 A GeV | Superheavy element physics and chemistry studies. <br><br> Studies of exotic nuclei including breakup and trapping. <br><br> In-beam gamma spectroscopy. <br><br> Dense plasma research. <br><br> High-density hadronic matter studies. | 1,300 | Facility for Antiproton and Ion Research (FAIR) is in progress, consisting of a superconducting double synchrotron, to be completed late in the decade. |

## TABLE 4.2 Continued

| Facility | Beam Characteristics | | Research Areas | Average Number of Users per Year | Future Upgrades |
| | Species | Energy | | | |
| --- | --- | --- | --- | --- | --- |
| Chinese Academy of Sciences (CAS) Heavy Ion Research Facility in Lanzhou (HIRFL) Lanzhou, China | Protons $^{12}C$ $^{238}U$ | 3.7 GeV 1.1 A GeV 520 A MeV | Superheavy element physics. Heavy ion and atomic physics. Cancer therapy. Accelerator physics and technology. | 200 | Large high-energy-density facility under consideration. Cancer therapy facilities including a booster, high current linear injector, and molecular injector are in planning. Production of radioactive beams. |
| J-PARC Tokai, Ibaraki, Japan | Protons Synchrotron | 30-50 GeV $>10^{14}$/s | Strangeness nuclear physics. Hadron physics. Neutrino physics with Kamioka facility. Kaon decay physics. Accelerator-driven nuclear waste transmutation. | 480 | Possibilities include improving the RF system and increasing the energy and intensity of the driver linac. Increasing the repetition rate of the facility. |
| Johannes Gutenberg Institute, MAMI Mainz, Germany | Electrons | 180-1,500 MeV | Electron scattering. Photon scattering. Studies of parity violation through electron scattering. X-ray generation. | 150 | A polarized "frozen spin" target is under way. |
| RIKEN, RIBF Wako, Saitama, Japan | Heavy ions $d \leq A \leq U$ | 345 A MeV | Superheavy element physics. Beams of exotic nuclei for nuclear structure and synthesis experiments. Nuclear astrophysics, element formation | 500 | A slow radioactive ion beam facility (SLOWRI) is under construction. Electron scattering facility for rare isotopes under way. |

continued

TABLE 4.2 Continued

| Facility | Beam Characteristics | | Research Areas | Average Number of Users per Year | Future Upgrades |
| | Species | Energy | | | |
|---|---|---|---|---|---|
| RIKEN, RIBF Wako, Saitama, Japan (*continued*) | | | | | New gas-filled separator for superheavy element physics studies in commissioning (GARIS-II). New heavy-ion linac for superheavy element experiments to run in parallel with other experiments. |
| RNCP Osaka, Japan | Protons Heavy ions A < 20 | 400 MeV 100 A MeV | Nuclear forces and mesons with proton beams. Quark and gluon properties. Neutrinos and dark matter. | 700 | Unknown at this time. |
| Tri-University Meson Facility (TRIUMF) Vancouver, British Columbia, Canada | Protons Radioactive ions | 500 MeV 6 A MeV | Nuclear astrophysics. Fundamental symmetries. Nuclear structure. Collaborations with CERN and Tokai on neutrino physics. | 600 | Deuterated scintillator array for neutron spectroscopy (DESCANT) in progress. IRiS solid hydrogen target system for ISAC is planned. GRIFFIN gamma-ray spectrometer in progress. |

SOURCE: Information contained herein is based on the IUPAP Worldwide Overview of Research Facilities in Nuclear Physics, Booklet 41, available online at http://www.triumf.info/hosted/iupap/icnp/Report41-final-12-07-11.pdf.

intended to be inclusive. A detailed description of the world's facilities has been assembled by Working Group 9 of the International Union of Pure and Applied Physics (IUPAP) and can be found at http://www.triumf.info/hosted/iupap/icnp/Report41-0317.pdf.

## UNITED STATES

The U.S. Department of Energy (DOE) supports three nuclear science user facilities—the Argonne Tandem Linac Accelerator System (ATLAS) at Argonne National Laboratory (ANL), the Continuous Electron Beam Accelerator Facility (CEBAF) at the Thomas Jefferson National Accelerator Facility (JLAB), and the Relativistic Heavy Ion Collider (RHIC) at Brookhaven National Laboratory (BNL)—and the National Science Foundation (NSF) supports one—the National Superconducting Cyclotron Laboratory (NSCL) at Michigan State University (MSU). The program at ATLAS is centered around particle beams from a superconducting linear accelerator that produces light and heavy ion beams to energies of around 15 A MeV. It is presently undergoing an upgrade to add accelerated fission fragments from a 1-curie californium source. The NSCL is a coupled superconducting cyclotron facility that produces heavy ion beams up to 200 A MeV that are used to produce rare isotope beams by fragmentation. These three facilities form the core of the experimental program in nuclear structure and nuclear astrophysics for the United States. Five smaller facilities that also focus on nuclear structure and astrophysics—three supported by DOE and two by NSF—are located at universities and one local DOE facility is supported at the Lawrence Berkeley National Laboratory (LBNL). The university facilities are at Florida State University, Notre Dame University, Texas A&M University, and the Triangle Universities Nuclear Laboratory. Also the NSF supports a physics frontier center—the Joint Institute for Nuclear Astrophysics (JINA)—as a consortium of three universities (Michigan State University, Notre Dame University, and the University of Chicago). The CEBAF facility features a continuous beam electron accelerator with energies up to 6 GeV and both polarized and unpolarized electron beams. The RHIC facility has two synchrotons used to produce counter rotating beams of heavy ions up to 100 A GeV and protons up to 250 GeV. At present, the beams undergo collisions at two different interaction points that are instrumented with the STAR and PHENIX detector systems. Table 4.1 summarizes the capabilities of the principal nuclear physics research facilities in the United States.

Each of the U.S. facilities has forefront research programs in nuclear science and all of them make significant contributions to the field. Both CEBAF and RHIC, which support user groups from many countries around the world, have made surprising scientific advances over the past decade. The recent discovery at RHIC from Au-Au collisions of a near perfect liquid form of hot matter has led to significant advances in the study of matter at high energy density. Using polarized proton beams, a program with strong support from Japan, RHIC has also provided data that put significant constraints on the contribution of gluons to the spin of the proton. The effort at RHIC for the coming decade will be dedicated to utilizing the upgrade luminosity to understand the quantum rules that govern the newly discovered form of matter and the proton spin. At CEBAF researchers are mapping, with high precision, the internal structure of the nucleon, and they are carrying out precision tests of the electroweak Standard Model. With the upgraded facility, researchers will extend the mapping of the nucleon internal structure and look for exotic particles that are predicted by quantum chromodynamics (QCD). The NSCL and ATLAS programs have large user communities in the United States and from abroad. As is pointed out in Chapter 2, both facilities have made significant discoveries in the past decade.

In low-energy nuclear physics the domestic stable and rare isotope beam facilities have enabled U.S. nuclear scientists to be among the leaders in advancing the field to further understand the emergence of order and collectivity from the chaotic assembly of protons and neutrons forming nuclei, and to determine how nature has accomplished the assembly of a wide range of different nuclides through chemical evolution. A particularly exciting direction that has emerged in this area is the study of unstable rare isotopes. As noted below, many countries around the world have made or are making significant investments in this area to provide accelerators to produce rare isotope beams. These rare isotopes open a new window on nuclear structure and have suggested that the concepts and paradigms developed from data with stable nuclei are often only a projection of a more general theory onto a small subset of nuclei whose only distinction is that they were the first to be studied. Studies focusing on all nuclei will lead to a more comprehensive theory of the nucleus. A next-generation rare isotope beam facility for the United States, FRIB, is now under construction at Michigan State University that will provide unprecedented access to a wide range of nuclei very far from stability (see Figure 2.3).

U.S. nuclear scientists have played a major role in developing neutrino physics into an important research thrust for nuclear and particle physics. In experiments dating back to the 1960s, nuclear scientists found that the number of neutrinos expected from the nuclear reactions taking place in the sun was significantly greater than what was observed. This long-unresolved solar neutrino puzzle was clarified with the discovery of neutrino oscillations, and the subsequent work that has shown just how complex the neutrino sector is. To date, neutrinos provide the only definitive indication of new physics beyond the Standard Model. Further study of neutrino properties, and neutrinos as cosmic messengers, could have transformative research results. U.S. nuclear scientists play active roles in many major efforts in neutrino physics, including experiments to determine the neutrino mass scale, to measure neutrinoless double-beta decay, to determine the precise mixing of neutrinos, and to study low-energy solar neutrinos.

Research in theoretical nuclear physics provides leadership and guidance to the experimental facilities, as well as supporting the existing experiments. Moreover, efforts are aimed at fundamental problems such as obtaining numerically exact solutions to the nuclear many-body problem, understanding the connection between QCD and nuclear physics, and predicting the existence of new phenomena. This research, which takes place on the international stage, offers opportunities for international collaboration between U.S. and foreign groups. In the United States, theory groups at the national laboratories and at universities work on a wide range of topics that cover the forefront areas of research in the field. The Institute for Nuclear Theory (INT) at the University of Washington serves as a national center for theoretical nuclear science research. Part of the INT program is to support workshops that bring together experimental and theoretical nuclear scientists from around the world to focus on important research topics. These workshops often have a major impact on both theoretical and experimental developments in the United States and other countries.

## EUROPE

The European Strategy Forum on Research Infrastructures list of future European large research infrastructures identifies two new facilities in the field of nuclear and hadron physics—the Facility for Antiproton and Ion Research (FAIR) to be built at the site of the GSI Helmholtzzentrum für Schwerionenforschung in Darmstadt, Germany, and SPIRAL2 at the Grand Accélérateur National d'Ions Lourds (GANIL) site in Caen, France. Both facilities have recently been funded and will spearhead research in fundamental and applied nuclear sciences in Europe after the completion of their first construction phases, which are expected to occur in 2016 and 2014, respectively. These facilities are complemented by a world-competitive experimental program in heavy-ion, radioactive ion-beam, and antiproton research at CERN and by a suite of national accelerator laboratories for lepton and hadron physics.

Europe supports several international centers for nuclear theory. The European Center for Theoretical Studies in Nuclear Theory and Related Areas (ECT*) in Trento, Italy, is focused on development of new theoretical approaches and connections of nuclear physics with astrophysics, elementary particle physics, and atomic physics. It hosts numerous workshops and research collaboration meetings, with good representation from the United States. Although it does not have a permanent faculty, it runs an annual doctoral training program for graduate students on a different specialized topic in nuclear theory each year, and has a strong postdoctoral research program.

The Helmholtz Alliance EMMI (Extreme Matter) Institute, a collaboration between 13 German and international partners, including the United States, has a strong theoretical effort in QCD, nuclear structure and reactions, and astrophysics, in particular exploiting interdisciplinary research such as strongly correlated systems.

The Jülich Supercomputing Centre (JSC) at the Forschungszentrum Jülich, with a number of supercomputers including the 1 Pflop JUGENE machine, is the principal European center for large-scale lattice QCD calculations in hadron physics and effective field theory calculations for nuclear structure. Large-scale astrophysical simulations are carried out at the Max Planck Institute for Astrophysics in Munich.

## FAIR and GSI

FAIR will be the next-generation facility for fundamental and applied research with antiproton and ion beams. It will provide world-unique accelerator and experimental facilities, allowing for a great variety of unprecedented forefront research efforts in physics and applied sciences. FAIR is an international project with 16 partner countries and more than 2,500 scientists and engineers involved in the planning and construction of the accelerators and associated experiments. FAIR will be realized stepwise. The Modularized Start Version will include a Heavy-Ion Synchrotron SIS100, an antiproton facility, and a Superconducting Fragment Separator and will contain experimental areas and novel detectors for atomic, hadron, heavy-ion, nuclear, and plasma physics, and applications in material sciences and biophysics. It is expected to be operational late in the decade. Completion of FAIR, including the synchrotron SIS300, is envisioned to follow thereafter. The existing GSI accelerator system, consisting of the UNILAC linear accelerator and the SIS18 synchrotron, will be used as the injectors to the FAIR accelerator complex. The GSI experimental storage ring ESR is planned to be available for experiments until construction of the new storage ring NESR begins after the completion of the Modularized Start Version.

FAIR research focuses on the structure and evolution of matter on both a microscopic and a cosmic scale, bringing our universe into one laboratory. In particular, FAIR will expand the knowledge in various scientific fields beyond current frontiers, addressing the following:

- The properties of the strong (nuclear) force and its roles in shaping the basic building blocks of the visible world around us and in the evolution of the universe;

- Tests of symmetries and predictions of the Standard Model, as well as the search for physics beyond it in the electroweak sector and in the domain of the strong interaction;
- The properties of matter under extreme conditions, at both the subatomic and the macroscopic scale of matter; and
- Applications of high-intensity, high-quality ion and antiproton beams in research areas that provide the basis for, or directly address, issues of applied sciences and technology.[1]

Compared to the existing GSI facilities, FAIR will provide an increase in beam intensities by factors of 100 to 10,000 and in beam energies by factors of 15-20. Moreover, the use of beam cooling techniques will enable the production of antiproton and ion beams of the highest quality—that is, with very precise energy and extremely small emittance. Upon completion, the FAIR accelerator complex can support up to five experimental programs simultaneously with beams of different ion species in parallel operation. This unique feature is made possible by an optimal balance in the use of accumulator, collector, and experimental storage rings.

The scientific user community of FAIR has organized itself into large international collaborations. The Compressed Baryonic Matter (CBM) collaboration will explore the phase diagram of hadronic matter by ultrarelativistic heavy-ion collisions; the Nuclear Structure, Astrophysics and Reactions (NuSTAR) collaboration will study the properties of exotic nuclei exploiting the unprecedented radioactive ion-beam capabilities of FAIR; and the antiProton ANnihilation at Darmstadt (PANDA) collaboration will use $\bar{p}$-p collisions to explore the role of QCD in hadron structure and dynamics. Beyond nuclear physics, the Atomic, Plasma Physics and Applications (APPA) collaboration will perform forefront research in atomic and plasma physics as well as in applied sciences like material research and in medical and biophysics.

FAIR builds on the existing infrastructure and experience at GSI. Research at the existing GSI accelerator complex will continue until FAIR becomes operational. Recent experimental highlights at GSI include the discovery of the heaviest elements from $Z = 107$-112 and the development of carbon beams to be used for radiation therapy in treating cancer.

## GANIL and SPIRAL2

GANIL in Caen, France, is an accelerator complex delivering both stable heavy-ion beams ranging from carbon to uranium and radioactive beams produced either

---

[1] Portions of this paragraph were adapted from C. Sturm, B. Sharkov, and H. Stöcker, 2010, 1,2,3 … FAIR!, *Nuclear Physics A* 834:682c.

in flight or with the Isotope Separation Online (ISOL) method in the Système de Production d'Ions Radioactifs Accélérés en Ligne (SPIRAL) facility. GANIL hosts a large suite of state-of-the-art detector systems to carry out a broad scientific program that focuses on properties of exotic nuclei and research in atomic physics, material sciences, and radiobiology.

The main scientific goal at SPIRAL2 is the exploitation of very intense radioactive ion beams to explore the properties of nuclides far from stability and to extend the knowledge of nuclear structure toward presently unexplored regions of the nuclear chart. SPIRAL2 will be an international facility. Already several memoranda of understanding with major laboratories, institutions, and ministries worldwide have been signed.

The driver of the SPIRAL2 facility is a high-power, continuous-wave superconducting linear accelerator (linac). It will accelerate a deuteron beam to produce neutrons that interact with a uranium target to produce radioactive ions by neutron-induced fission. The facility will yield radioactive ion beams in the mass range A = 60 to 140, with intensities that will be unique in the world for some species. These beams will be available at energies ranging from a few keV/A at the new experimental hall for low-energy exotic nuclei up to 20 A MeV at the existing GANIL experimental areas, where a suite of next-generation detectors will be used to detect gamma-rays, charged particles, and neutrons.

The SPIRAL2 linac will also accelerate high-intensity heavy-ion beams up to 14.5 A MeV. These heavy ion beams will be used to produce neutron-deficient nuclei by the Isotope Separation On-Line (ISOL) method or very heavy nuclei by fusion evaporation. Another option, to produce very heavy or neutron-deficient nuclei, will exploit the in-flight method with the Super Separator Spectrometer ($S^3$). The high neutron flux produced with the deuteron beams at the new Neutron for Science Facility (NFS) will provide additional experimental opportunities for applied research.

The timeline of SPIRAL2 anticipates commissioning of the linear accelerator, the $S^3$ spectrometer, and the NFS experimental hall in 2012, with the commissioning of radioactive ion-beam production and the DESIR low-energy facility to follow in 2014. In the future, GANIL plans to increase the intensity of medium- and heavy-mass radioactive ion beams by adding a second heavy-ion injector. The SPIRAL2 project is seen as an important step towards EURISOL, a large-scale ISOL facility for Europe, for which a conceptual design study has been carried out within the fifth framework program of the European Union.

## CERN

A Large Ion Collider Experiment (ALICE) is the largest nuclear physics experiment in the world. It exploits the physics potential of nucleus-nucleus collisions at the multi-TeV energy scale offered by the LHC at CERN. To reach this goal, the ALICE Collaboration of more than 1,000 members from 116 institutions in 33 countries has built a multipurpose heavy-ion detector.

The physics motivation is to study strongly interacting matter at extreme energy densities, and the quark gluon plasma in particular. This new phase of matter and its properties are key issues in QCD for understanding the fundamental phenomena of quark confinement and chiral-symmetry restoration. For this purpose, a comprehensive study of the hadrons, electrons, muons, and photons produced in the heavy-ion collisions is required, which can be achieved by various dedicated components of the ALICE detector. The experiment started operation in 2010 with proton-proton collisions. A first successful heavy-ion run with Pb-Pb collisions was performed at the end of 2010.

The full experimental program of ALICE will take place over more than a decade and will benefit from the successive upgrade of the beam energies available at the LHC. For the mid-term, an ambitious upgrade of the detector is envisioned, which will include a new inner tracker to extend the Cherenkov and calorimeter coverage and to increase the rate capability for physics with rare probes and the construction of a new tracking calorimeter system that will operate at forward angles relative to the beam direction.

Other LHC collaborations have significant efforts aimed at searching for new physics in both p-p and Pb-Pb scattering. Two of these—ATLAS and the Compact Muon Solenoid (CMS) experiment—have significant participation from the U.S. relativistic heavy-ion physics community. The primary goals of the heavy-ion programs for ATLAS and CMS are similar to those for the ALICE collaboration. Each collaboration uses the physics variables that their detector is optimized to measure to study the physics in these ultra-high-energy heavy-ion collisions.

The Isotope Separator Online Detector (ISOLDE) is primarily a nuclear physics facility at CERN that produces radioactive beams through fission, spallation, and fragmentation reactions induced by 1.4-GeV protons from the CERN Proton Synchrotron Booster for research in nuclear structure, nuclear astrophysics, and fundamental physics. With several decades of accumulated experience in target and ion-source knowledge, ISOLDE has extracted and separated about 700 different isotopes of more than 70 elements, which is by far the largest number of isotopes to be available for users at any ISOL facility in the world. With the installation of the postaccelerator REX-ISOLDE, it is now possible to accelerate radioactive isotopes up to mass number 238 to energies of 3 A MeV. This energy will be increased to 10 A MeV through the HIE-ISOLDE (High Intensity and Energy) project. Additionally, the project will improve beam quality and intensity. These goals will be achieved by replacing the REX linac by superconducting cavities and by the higher intensity primary beams provided by the new CERN injector LINAC4. HIE-ISOLDE has been approved by the CERN Research Board and is scheduled to be available to users around 2014.

For the last decade the Antiproton Decelerator (AD) at CERN has been providing low-energy antiprotons for experiments to produce and study cold antihydrogen atoms, to investigate the spectroscopy of antiprotonic helium, and to study the biological effects of antiprotons on living tissue. The physics motivations for these experiments are tests for charge conjugation-parity-time symmetry (CPT) and the determination of fundamental constants like the electron to antiproton mass ratio and the antiproton magnetic moment. In the future an approved experiment, the Antihydrogen Experiment: Gravity, Interferometry, Spectroscopy (AEGIS), aims at measuring the gravitational coupling of matter and antimatter.

The approved experimental program at the AD extends to 2016. As an addition to the AD, a further deceleration stage, the Extra Low Energy Antiproton Ring (ELENA) project, is currently under discussion. The ELENA project would bring antiprotons to the keV energy range and would bridge the gap until the Facility for Low-Energy Antiproton and Ion Research (FLAIR) project at FAIR, which is not part of the Modularized Start Version, becomes available.

### Other European Facilities

The physics community in Europe also operates a host of smaller hadron and lepton accelerator facilities. These laboratories run focused programs in which they are quite competitive on the international scale, and all of them have significant theory groups.

The hadron beam facilities include the two national laboratories in Italy, at Legnaro and Catania, the Accelerator Laboratory at the University of Jyväskylä in Finland, the Kernfysisch Versneller Instituut (KVI) at Groningen in The Netherlands, the cooler synchrotron (COSY) facility at the Forschungszentrum Jülich in Germany, the Linear Accelerator Near the Tandem of Orsay (ALTO) facility in Orsay, France, and the facilities at the Joint Institute for Nuclear Research in Dubna, Russia. Besides performing dedicated research in nuclear science and its applications, these facilities and a suite of smaller laboratories, often located at universities, play an important role in educating the next generation of nuclear scientists and in the development and testing of devices to be used at the large-scale facilities.

The lepton beam facilities include the Microtron Accelerator for X-rays (MAX) IV Laboratory (MAX-lab) in Lund, Sweden; the two electron accelerators, Mainz Microtron (MAMI) at the Johannes Gutenberg University of Mainz and Electron Stretcher and Accelerator (ELSA) at the University of Bonn, in Germany; the high-energy physics experiment Common Muon Proton Apparatus for Structure and Spectroscopy (COMPASS) at the Super Proton Synchroton at CERN; the Frascati National Laboratory in Italy with its $e^+e^-$ meson facility, Daphne; and the

$e^+e^-$ VEPP facility at the Budker Institute of Physics in Novosibirsk, Russia. Finally, the Technische Universität Darmstadt operates a superconducting low-energy electron accelerator, the superconducting Darmstadt electron linear accelerator (S-DALINAC).

Many of these hadron and lepton beam facilities are considering dedicated upgrade programs. These as well as detailed descriptions of the facilities and their physics goals and achievements can be found in the report *Perspectives of Nuclear Physics in Europe,* published by the Nuclear Physics European Collaboration Committee (NuPECC) in December 2010.[2]

The Extreme Light Infrastructure (ELI-NP) is a proposed high-energy laser research facility of the European Union to be constructed by 2016 in Bucharest, Romania. Based on a linear electron accelerator, the Doppler shift of Compton-backscattered laser photons off relativistic electrons is used to generate a high-energy gamma-ray beam for basic research in nuclear structure and applications. The average photon flux at ELI-NP is envisaged to be similar to that of the next-generation laser Compton backscattering facility, the mono-energetic gamma ray facility (MEGa-Ray) at Lawrence Livermore National Laboratory (LLNL) (operations planned for 2013), and will exceed the flux at existing facilities by several orders of magnitude.

Europe has had a long tradition in underground science with the development in the 1980s of the Gran Sasso facility. It is still the largest underground laboratory in the world and houses detectors carrying out a wide range of fundamental studies. It also has served as the home for the Laboratory for Underground Nuclear Physics (LUNA) facility, which has a low-energy underground accelerator and detector setup to measure nuclear reactions of astrophysical importance. The facility is limited in reach by the accelerator and cannot access reactions such as the very important alpha-induced reactions.

## ASIA, AFRICA, AND AUSTRALIA

Over the past decade, China and Japan have made major investments in new nuclear physics facilities, which have led to a substantial growth of the field in those countries. The buildup of nuclear physics capabilities in Asia appears to be continuing into the future with rapid growth of funding in India and for a major new nuclear physics facility to be built in South Korea. Groups from China, India, and Japan have been playing major roles in the two large collaborations at RHIC. Many of them also are involved in the heavy-ion program at the LHC. Groups

---

[2] European Science Foundation, 2010, Perspectives of Nuclear Physics in Europe-NuPECC Long Range Plan 2010. Available online at http://www.nupecc.org/pub/lrp10/lrp2010_final_hires.pdf; last accessed on October 27, 2011.

from Japan also have been carrying out experiments that focus on hyperon physics at JLAB. For many years, U.S. nuclear physicists have worked with Japanese colleagues on neutrino experiments at Kamioka. Recently, a new reactor neutrino experiment began in China that includes nuclear and high-energy physicists from the United States.

The two main streams of nuclear theory research in Japan continue the Japanese tradition of excellence in many-body theory. The first area, the many-body physics of quarks and gluons, is very strong and growing as a consequence of experimental programs at RHIC and the Japan Proton Accelerator Research Complex (J-PARC). The RIKEN-BNL Research Center (RBRC) headquartered at BNL, has been training young scientists from the United States and Japan who are now leaders in the field. The second area, the many-body physics of hadronic systems (nucleons and hyperons), is now regaining interest after losing young people to QCD studies, in part owing to the Radioactive Ion Beam Facility (RIBF) at RIKEN, near Tokyo. The focus has been on first-principles calculations, including massive lattice QCD and ab initio many-body calculations. A significant factor in this work has been the very strong lattice QCD group at the University of Tsukuba, which has stimulated interest and abilities in this area.

Japan has completed construction of two large-scale projects for nuclear physics during this decade. RIBF is now providing particle beams from hydrogen to uranium at up to 350 A MeV. Through extensive work on beam development, beam intensities for some light-ion beams have now reached the design goal. During the first several years of operations, intensities of the heaviest beams have been limited by the lifetime of stripper foils. A gas stripper system, which has been developed for the facility and should be operational by 2012, will allow for significant increases in the intensity of heavy beams. Even at the present level of operation, RIBF has the highest intensities in the world for very neutron-rich secondary beams.

J-PARC, a large new facility near Tokai, Japan, has been built to accommodate a very broad physics program ranging from neutron scattering for materials science to high-energy neutrino-scattering experiments. The accelerator complex includes a high-beam power linac followed by a 3-GeV synchrotron, which is designed to deliver up to 1 MW of protons, and a 50-GeV synchrotron, which is designed for a proton beam power of 0.75 MW. Ground breaking for the facility was in 2002 and commissioning of the linac began in 2006. The first neutrino beam from the 50-GeV synchrotron to the neutrino detector at Kamioka occurred in 2009. Secondary beams of muons, pions, kaons, and antiprotons will be used to carry out a broad hadronic and fundamental symmetries physics program. Commissioning of the first phase of the hadron physics experimental area began in late 2009 and will continue as spectrometers and beam lines are completed. The beam power at J-PARC is projected to increase from present values of about 200 kW and reach design goals by late 2013. Recently the theory center at the High Energy Accelerator

Research Organization (KEK) in Japan at the request of the Japanese Nuclear Theory Association, appointed five university nuclear theorists as visiting scientists to help organize J-PARC theoretical activities. The massive Tohoku earthquake that struck Japan on March 11, 2011, caused considerable damage to the J-PARC facility. However, most of the damage was limited to the surface infrastructure; fortunately, the accelerator components suffered only moderate damage. Following an assessment of the facility, repairs were carried out to have the facility back on line for beam tests in the first quarter of 2012. This unfortunate event will probably impact the facility's plan to reach the full design intensity by late 2013.

In addition to the two international facilities for nuclear physics in Japan, several smaller facilities continue to operate. Among the largest of these is the Research Center for Nuclear Physics (RCNP) in Osaka. The facility is built around an azimuthally varying field (AVF) and a Ring cyclotron. These accelerators provide stable light-ion beams from H up to ions with $A < 20$. The traditional focus of the program has been on high-resolution studies of spin and isospin degrees of freedom. As a national research center RCNP is primarily used by Japanese experimentalists. Typically about 15 percent of the users come from outside the country.

Supercomputing in Japan provides considerable support to the nuclear theory effort, well beyond that available in the United States. The national 10 Pflop (from 2012) supercomputer KEI, at the Advanced Institute for Computational Science in Kobe, is one of the five "urgent major projects" in science and technology; it supports a unique collaboration among nuclear, particle, and astrophysics scientists and is supporting a growing number of theory positions. Formidable supercomputing power is available to nuclear physics through the KEK supercomputers, with over 1 Pflop; the Center for Computation Sciences at the University of Tsukuba, with 95 Tflop; and the Yukawa Institute for Theoretical Physics (YITP) at the University of Kyoto, with 91 Tflop. In addition to nuclear theory efforts, Japan supports a growing theory collaboration among string theorists, condensed matter theorists, and nuclear theorists (who play a joining role among these disparate fields) that is centered on the physics of strongly coupled many-body systems.

Japan supports two important programs with the United States in nuclear theory: the RBRC, with its focus on QCD, and the Japan-U.S. Theory Institute for Physics with Exotic Nuclei (JUSTIPEN) at RIKEN, which focuses on the physics of rare isotope beams. These two programs have been very successful in developing collaborations between theorists in the two countries. Following on the success of JUSTIPEN, a new initiative has been launched—the French-U.S. Theory Institute for Physics with Exotic Nuclei (FUSTIPEN). Like its counterpart, the program at FUSTIPEN also will focus on physics with rare isotope beams. Recently a new Japan-U.S. Institute for Physics with Exotic Nuclei (JUSEIPEN) was set up to promote experimental activities by U.S. participants at the RIKEN facility RIBF.

China is promoting nuclear physics research through investments at facilities in Lanzhou and Beijing. At Lanzhou, a separated sector cyclotron is used to provide beams for acceleration and storage in a synchrotron ring that can accelerate heavy ion beams up to 1 A GeV for carbon-12 and 500 A MeV for uranium-238. Primary beams from the synchrotron can be used for experiments. They also are used to produce radioactive beams through fragmentation that can be collected in a second storage ring at the end of a fragment separator beam line, where they are cooled and accelerated. Both storage rings are equipped with internal target stations. Commissioning of the main storage ring began in 2005 and was followed by commissioning of the second storage ring. The facility is now in operation and carrying out experiments. It is used primarily by scientists in China.

The Beijing rare ion beam facility, BRIF, is an upgrade of the existing nuclear physics laboratory at the Atomic Energy Commission in Beijing. The central part of the upgrade is the construction of a high-current 100 MeV compact proton cyclotron. The proton beam will be used to produce secondary radioactive beams by the ISOL technique, which will be accelerated in the existing tandem accelerator. A superconducting linac will be added after the tandem for further beam acceleration. Design and construction of the cyclotron began in 2006. Civil construction for building modifications is scheduled to begin in late 2010. The ISOL system and the superconducting linac designs have been completed and construction is now under way. Like the facility at Lanzhou, BRIF will be used primarily by researchers in China.

Experiments at the new facilities in China drive theoretical research that focuses on nuclear reactions, nucleosynthesis, nucleon-nucleon interactions, QCD matter in heavy ion collisions, and hadronic physics. Work is driven in part by experimentation at newer facilities in China, such as the Cooling Storage Ring of Heavy Ion Research Facility in Lanzhou (HIRFL-CSR), the Beijing Spectrometer III (BES-III) at the Beijing Electron Position Collider II (BEPCII), and the Shanghai Synchrotron Radiation Facility (SSRF).

Rather recently, China built an underground laboratory to carry out a broad research program in fundamental physics at Jinping. The facility is built under a large mountain and has a tunnel for drive-in access. When completed, it will have about the same space for experiments as the Gran Sasso facility in Europe but at twice the depth.

Today, the nuclear theory effort in China is considerably smaller than the one in Japan, with most of the nuclear theorists at the Institute of Nuclear Science, the Institute of High-Energy Physics of the Chinese Academy of Sciences, the Center for Theoretical Physics, Peking University, and Tsinghua University, all in Beijing, as well as Huazhong Normal University in Wuhan and Fudan University in Shanghai. The rapid growth of Chinese university research and ever-increasing contact with foreign research—including support of active theory collaborations with theorists in the United States, Germany, and Japan, and experimentalists at major foreign laboratories, e.g., JLab and RHIC in the United States, LHC and GSI in Europe, and RIKEN in Japan—provides a framework for an expanded program in nuclear theory in China.

As part of the effort to grow its program in nuclear physics, China has created a significant number of new faculty positions at its major research universities. The universities are aggressively recruiting faculty who were originally from China but have been trained and employed in the United States and Europe. This enhances the Chinese research program at the expense of other efforts such as the U.S. effort.

India has started making a significant investment in nuclear physics, which is driven in part by the anticipated growth in the use of nuclear power for the region. As part of this effort, the laboratory at Kolkata is undergoing an upgrade to radioactive beams. Two different approaches are being developed to produce secondary beams. One approach uses beams from the K130 cyclotron on a thick target with an ISOL type system. The other approach uses photon-induced fission on uranium. A 50-MeV high-power electron linac is being built in cooperation with TRIUMF to produce the photons for fission. After separation, relatively low-energy radioactive ion beams (RIBs) will be produced by acceleration through a room-temperature linac. Other nuclear physics laboratories in India are carrying out infrastructure improvements and developing new detector technologies. Beyond the nuclear physics laboratories, India has also begun developing a deep underground laboratory for neutrino studies (INO).

In a very recent development, South Korea is planning to construct a major new international rare isotope accelerator facility for nuclear physics, KoRia, which will be part of a new Basic Science Institute (BSI) to be built in the country. Legislation authorizing the construction of a new city, the home of the BSI, has now been passed in South Korea. Planning for the nuclear physics facility has been under way for about a year. The concept that is being developed includes a 70-kW beam of protons up to 100 MeV that will be used for ISOL production of secondary particles. A superconducting linac will accelerate the ISOL beams to typical energies of about 15 A MeV for reaction studies. A second superconducting linac will be added to the facility to boost ion beams to about 200 A MeV for uranium. The higher energy linac will be built to accelerate both RIBs from the first part of the facility and high-intensity stable beams from a superconducting linac injector. Following the high energy linac, a fragmentation beam line will be set up to collect radioactive ions produced either from the accelerated RIBs or stable beams. As part of the project, spectrometers are being designed to use both the 15 A MeV RIBs and the RIBS produced either by fragmentation or direct acceleration at high energies. The laboratory being developed will serve as a catalyst to increase the nuclear physics workforce in South Korea and it also will be developed as an international user facility.

Korea presently supports several cooperative national research efforts, e.g., between the World Class University (WCU) project at Hanyang University in Seoul and the Asian Pacific Center for Theoretical Physics (APCTP) at Postech in Pohang. The focus here is to develop theoretical study and awareness of interdisciplinary problems between particle, hadronic, and nuclear physics, together with astrophysics and condensed matter, with connections to future experiments at FAIR, J-PARC, and rare isotope accelerators, as well neutron stars. Also supported is the multiinstitutional heavy ion meeting (HIM) effort in heavy ion physics. International collaborations in nuclear theory include that of the APCTP with a number of international centers, WCU with the Yukawa Institute for Theoretical Physics in Kyoto, and HIM with theorists at CERN, GSI, and RHIC.

In Africa, the major research facility for nuclear science is the iThemba laboratory in South Africa. A separated sector cyclotron operated at the laboratory near Capetown is used for isotope production and for research in nuclear science. Beyond this, several other countries operate small research facilities that are used in part for nuclear science.

Accelerator-based nuclear physics research in Australia is centered at the Australian National University in Canberra, Australia. The science program is built around a Pelletron electrostatic accelerator and a superconducting linac that together provide a wide variety of particle beams. An upgrade project began in 2009 to improve the system reliability and to develop the capability to produce radioactive ion beams.

## CANADA AND LATIN AMERICA

Nuclear science in Canada has a rich history that dates from Rutherford's early work at McGill University. Today it is centered at the TRIUMF laboratory in Vancouver, Canada. Over the last decade, TRIUMF has developed an ISOL-based program of radioactive beam production. The 500-MeV proton cyclotron serves as the driver for producing secondary beams. With a primary beam power of nearly 100 kW, TRIUMF has produced the highest intensities of many light ion beams in the world. A superconducting linac serves to accelerate the RIBs to energies of 10 A MeV for light species. TRIUMF and Canada have made a significant investment in ancillary equipment for studying isotopes (e.g., the ISAC-I and ISAC-II experimental halls and their associated detectors ), making it one of the top RIB facilities in the world. Nuclear theory in Canada is supported at TRIUMF and at several universities throughout the country.

Building on its expertise in isotopes for nuclear physics, TRIUMF is also pursuing a program in nuclear medicine, using proton beams for treatment of ocular melanomas and studying the physics, chemistry, and biology of medical isotopes for diagnosis and treatment of cancer and neurodegenerative disease. Together with commercial partner Nordion, Inc., the team produces about 2.5 million patient doses of medical isotopes every year that are exported around the world.

Canada has ambitions to further expand its programs in isotopes for physics and medicine. Funding was approved recently to develop the Advanced Rare Isotope Laboratory (ARIEL) at TRIUMF, a facility that will feature a 0.5-MW superconducting radio frequency electron linear accelerator and two additional isotope-production targets. ARIEL will employ photofission to complement the cyclotron's proton-spallation technique; photofission promises access to neutron-rich isotopes at intensities beyond what are presently available. ARIEL will generate its first beams by 2015. In collaboration with a group from RCNP in Japan, TRIUMF plans to develop a source of ultracold neutrons and use it to carry out tests of fundamental symmetries such as a measurement of the neutron electric dipole moment.

Canada has been a world leader in underground science for many years. Following the successful completion of the solar neutrino experiment, SNO, at the Sudbury mine, Canada has been investing in expanding its underground program. The SnoLAB facility in Canada is now one of the deepest sites in the world. A broad research program has been established there, which will continue to make Canada an international leader in the field.

Nuclear science has a long tradition in Latin America, where research programs were initiated more than a half century ago. Today active research programs exist in most South American countries and Mexico, together with a large fraction of activities directed at important pressing problems where nuclear physics can contribute, including health, the environment, and energy. Argentina, Brazil, and Mexico have broader programs than other countries in the region, with operating facilities for basic research and educational programs that offer doctoral degrees in basic and applied nuclear physics. The research and educational activities in Venezuela, Colombia, Peru, and Chile are on smaller scales but are nonetheless significant within their respective country's basic research portfolios.

Argentina's nuclear physics research activities involve both basic research and applications. The National Atomic Energy Commission (CNEA), a government entity that maintains major research centers spread out throughout the country and that currently is in a very active expansion phase, plays a significant role. In Buenos Aires the major facility for research is the 20-MV Tandem (TANDAR), which has a basic research program centered on the investigation of nuclear reactions induced by beams of stable, weakly bound nuclei. The accelerator includes a Q2D magnetic spectrometer, a microbeam facility (beam spots of about $1 \mu^2$) with high resolution X-ray detection, an external beam irradiation facility with online dose determination, and heavy-ion identification based on a time-of-flight facility. The other major CNEA centers are in Bariloche and Ezeiza, each with unique capabilities for basic and applied work. Universities also play important roles in the use of CNEA facilities and in the development of and participation on other external activities like the Pierre Auger Observatory. Argentina has also developed strong programs in medical physics and nuclear engineering.

Nuclear physics research in Brazil and Mexico is conducted at a large number of universities with programs in basic research and applied nuclear physics. In both countries nuclear scientists also play significant roles in international collaborations (Auger, RHIC, LHC-CERN) and in the development and implementation of nuclear science applications for their own use. In Brazil one of the major investments is a new radioactive ion beam facility (RIBRAS) coupled to the 8-MV Pelletron Tandem at the University of Sao Paulo. RIBRAS is dedicated to basic research and to the training of nuclear scientists. In Mexico nuclear physics is at the core of activities in radiation physics, medical physics, and cosmic rays.

The existing political climate is presently favorable for developing collaborations in Latin America (MERCOSUR and other initiatives). Nuclear scientists in the region are helping to take advantage of this situation and join forces in projects of common interest. An example is the recent proposal for ANDES. The design and development of ANDESlab, an underground laboratory in Argentina, is a joint South American effort between Argentina, Brazil, and Chile. The ANDESlab is foreseen as a complete underground laboratory to enable the measurements of neutrinoless double-beta decay, neutrino oscillations, and nuclear reactions of importance to stellar evolution. Table 4.2 summarizes the capabilities of the principal nuclear physics research facilities outside the United States.

## U.S. NUCLEAR SCIENCE LEADERSHIP IN THE G-20

Large investments in new facilities for nuclear science have been made over the last decade by G-20 member nations and more are planned in this coming decade. While it is difficult to get a cost for these facilities, a conservative estimate is that over $4 billion will be spent by other G-20 nations before 2020. U.S. researchers are taking advantage of this by establishing research collaborations with local groups at the new and upgraded facilities. But if the United States is to remain a global leader in nuclear science, it must proceed with the plans that it has developed for its own new and upgraded facilities.

Through the long-range planning process of the Nuclear Science Advisory Committee (NSAC), a path forward for U.S. nuclear science has been laid out that would provide world-class facilities for research in several parts of the field well into the future. Over the past few decades, nuclear science around the world has evolved from a discipline where experiments were mostly carried out at local facilities to one that depends heavily on large national user facilities. During this time, the United States developed two major facilities, RHIC at BNL and CEBAF at JLAB.

Because the present energy capability of CEBAF limits the research program, the facility is undergoing an energy upgrade from 6 to 12 GeV. With the upgraded facility, the United States is expected to lead this area of hadron physics research well into the next decade.

In ultrarelativistic heavy-ion physics, the United States will share leadership with the LHC at CERN, where for one month of each year the formation of matter at high energies and densities will be studied in Pb-Pb collisions with more than 20 times the collision energy and perhaps twice the temperature of collisions at RHIC. These experiments will yield the highest energy probes of quark-gluon plasma. U.S. scientists and students are significant partners in this European effort, building components and developing software for the ALICE, ATLAS, and CMS detectors to be used in these studies. Meanwhile at RHIC, detector and luminosity upgrades are proceeding that, together with the flexibility to vary collision systems and energies over a wide range, will give RHIC researchers many advantages in the systematic investigation of the properties of quark-gluon plasma in varied regimes.

A major research focus for nuclear science worldwide lies in the development of facilities to produce and study the exotic nuclei that nature makes during catastrophic events such as a supernova. The science motivating these studies is to understand the emergence of order or collectivity from chaos and the evolution of stars, which produce the elements and lead, ultimately, to the chemical evolutions that allow the development of life. The United States plans to build the Facility for Rare Isotope Beams (FRIB) at Michigan State University, which will be a world-leading facility in rare isotope science when it is completed in 2018. For many rare isotope beams, FRIB will provide the highest beam intensities available in the world. It also is expected to be the only facility that will provide low-energy reaccelerated beams produced via projectile fragmentation. Here, however, it will have major competition from other facilities around the world. Today, the United States holds a leading position in the development of tools and techniques to study nuclear astrophysics. The establishment of the physics frontier center, JINA, has assembled an exemplary combination of astronomers, astrophysicists, nuclear theorists, and nuclear experimentalists. Together, they are expected to more rapidly bring about an understanding of stellar evolution as well as the explosive cosmic events that contribute to the synthesis of matter. JINA has served as a model for the Helmholtz Foundation in Germany, which has developed several similar centers of excellence, and for the formation of a Brazilian center for nuclear astrophysics. India, China, and Argentina have sought assistance from experts at JINA as they develop accelerators to be deployed in underground laboratories.

While U.S. scientists were pioneers in the studies of neutrinos, some of the most definitive work to date has been done at underground laboratories in Japan and Canada, with United States participation. There is great interest worldwide in understanding neutrino properties and measuring neutrinoless double-beta decay, which has resulted in very rapid development of underground laboratories at several locations around the world in a very short time. If the United States proceeds with plans to develop an underground science laboratory, it can become a world leader in this field. Fitting naturally into a new underground laboratory would be the U.S.-led Majorana experiment to measure neutrinoless double-beta decay and an accelerator laboratory for low-energy nuclear astrophysics, which would further unravel the details of stellar evolution.

Over the years, U.S. nuclear scientists have been innovative in exploiting opportunities to carry out experiments at facilities around the world. This effort should be commended and encouraged in the future. But without a significant U.S. investment in facilities throughout this decade, which includes the operation of existing facilities, the timely construction of new capabilities, and the flexibility to take advantage of new answers and discoveries, the United States could quickly lose its leadership in many of the forefront areas of physics research.

## Highlight: The Fukushima Event—A Nuclear Detective Story

On March 11, 2011, a massive earthquake and tsunami struck Japan. The Fukushima Dai-ichi reactor complex suffered severe damage and experienced a series of explosions in subsequent days (Figure FUK 1). Radioactivity was released into the atmosphere and deposited on large areas of Japan (Figure FUK 2). In the United States there was concern that this radioactivity would cross the Pacific Ocean and present a health hazard in the United States as well. Physicists in the nuclear physics community working on neutrinoless double beta decay realized that the radioactivity might render the backgrounds too high for the success of their experiment. At the University of Washington an air-monitoring system was quickly set up to collect and measure airborne radioactivity attached to dust particles. Scientists at Berkeley prepared to monitor rainwater for fallout. Across the country, other university nuclear physics groups set up detectors and collection methods, and the universities became one of the main sources of detailed information about the radioactivity.

Radioactivity associated with the Fukushima reactors arrived at Seattle during the night of March 17 (PST). The strongest signal was from iodine-131, and gamma-ray lines from tellurium-132, cesium-134, and cesium-137 were seen as well. The 8-day iodine-131 and 3.2-day tellurium-132 provided clear evidence that the primary source of the radioactivity was a reactor that had recently been in operation, not spent fuel. However the 21-hour iodine-133 was not seen, giving assurance that the nuclear chain reaction had been successfully shut down at about the time of the earthquake (Figure FUK 3). The presence of cesium-134 is typical of reactor debris as distinct from nuclear weapons, because it is blocked by stable xenon-134 from being produced in fission and can only be produced by neutron capture. While superficially similar to the Chernobyl disaster, the Fukushima radioactivity inventory is very different, with only three isotopes present in abundance instead of the broad spectrum of fission products released at Chernobyl. One possible explanation for this is the release of contaminated steam instead of burning fuel elements, as happened at Chernobyl. In addition to gaining detailed insight into the unfolding disaster in Fukushima, the university groups were able to reassure the public that no dangerous levels of activity were reaching the United States.

The efforts at the University of Washington were only part of the efforts that nuclear scientists across the United States dedicated to helping assess and address the challenges in Japan as well as to communicate with the public. As an example, one of the NNSA's radiological field monitoring team leaders was Daniel Blumenthal. The goal of his team was to provide timely information on environmental conditions so that fact-based protective action decisions could be made by decision makers that included the U.S. Armed Forces, the U.S. Embassy in Tokyo, and the Government of Japan. The members of this interdisciplinary team were prepared to perform "in extremis" science, where there is no time for lengthy deliberation or peer review yet accuracy during both collection and analysis is paramount. Dr. Blumenthal's 33-member team was deployed to Japan via a dedicated U.S. military airlift on March 14 at the direction of the White House. They brought radiation detection systems to perform both aerial and ground-based radiological surveys. The aerial systems were mounted on Air Force aircraft and operated by the deployed scientists and technicians. The ground-based systems included equipment for basic exposure rate measurements; high-resolution in situ gamma-ray spectroscopy; and sampling of air and soil. In addition to the field team, an essential component of the response capability was a home team staffed by a similar array of nuclear scientists and other experts trained to analyze field data quickly and to communicate their results to leaders in the field and in Washington, D.C.

One of the biggest challenges during the Japan response was translating the measurements into actionable information. It was critical that the information be framed so that it made sense

FIGURE FUK 1 Aerial image of the Fukushima nuclear power plant complex after the initial explosions in March 2011. SOURCE: Copyright Air Photo Service Co. Ltd., Japan.

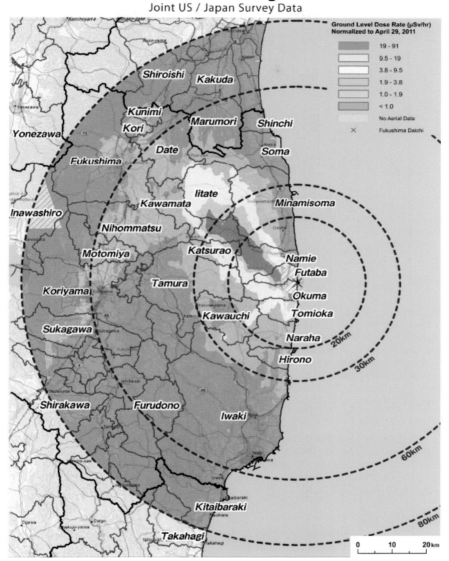

# Aerial Measuring Results
### Joint US / Japan Survey Data

FIGURE FUK 2 Example of aerial measurements of the radiation plume in the region near the Fukushima nuclear power plants. SOURCE: Joint United States-Japan survey data provided by the U.S. Department of Energy, National Nuclear Security Administration (NNSA), Office of Emergency Response.

*continued*

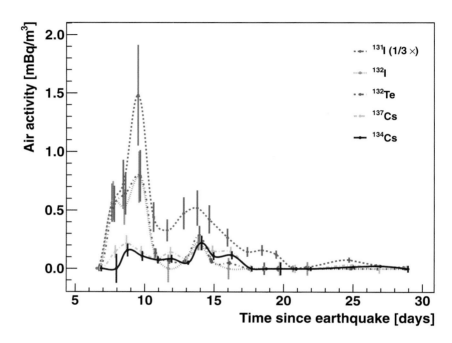

FIGURE FUK 3 Airborne activity observed in Seattle for radioactive isotopes of iodine, tellurium, and cesium in millibecquerels per cubic meter. As a reference, the average human body is exposed to 4,400 becquerels from the decay of potassium-40, a naturally occurring radioactive isotope of potassium. SOURCE: Reprinted from J. Diaz Leon, D.A. Jaffe, J. Kaspar, A. Knecht, M.L. Miller, R.G.H. Robertson, A.G. Schubert, 2011, Arrival time and magnitude of airborne fission products from the Fukushima, Japan, reactor incident as measured in Seattle, WA, USA, *Journal of Environmental Radioactivity* 102 (11): 032, Copyright 2011, with permission from Elsevier.

and that the decision makers would embrace the scientific process and have confidence in the information provided. In Japan, the success of the effort was endued by the fact that both civilian and military leaders requested that the technical experts remain in Japan as long as possible. Dr. Blumenthal communicated the results to a wide range of audiences, including to the general public on ABC News.[1]

---

[1] Available at http://abcnews.go.com/International/video/tracking-japan-radiation-from-the-sky-13628048.

# 5

# Nuclear Science Going Forward

## WAYS OF MAKING DECISIONS

### The Long-Range Plan Process

For over three decades, priorities for future U.S. nuclear physics facilities have been developed through a long-range planning process organized through the Nuclear Science Advisory Committee (NSAC), in response to requests for guidance from the Department of Energy (DOE) and the National Science Foundation (NSF). NSAC was initiated in 1977 as a joint advisory committee to NSF and DOE and currently is chartered under the Federal Advisory Committee Act (FACA). Lead responsibility for NSAC's direction, including the selection of its members and the development of meeting agendas and charges for the committee, are shared by DOE and NSF.[1] The U.S. nuclear physics community devotes substantial effort to its long-range planning processes, with several hundred scientists participating actively. The plans that result reflect the broad and varied scientific perspectives that are a strength of nuclear physics. The coordinating role that NSAC plays is valuable as is its role as a vehicle for transmitting the input and guidance that comes from the community to the priority-setting processes in nuclear physics.

The first comprehensive Long-Range Plans (LRPs) for nuclear science were published by NSAC in 1979 and 1983. Following on their success, DOE and NSF have asked NSAC to develop new LRPs every 5 to 7 years. As the LRP planning

---

[1] NSAC Web site; http://science.energy.gov/np/nsac/. Last accessed August 26, 2011.

process evolved, the general nuclear physics community has become more involved. These LRPs provide a view of the significant scientific opportunities in the field and furnish priorities and recommendations for realizing them.

In many cases these recommendations have advocated and led to the construction or completion of a major facility. Indeed, the two large DOE-funded existing U.S. facilities for nuclear physics, the Continuous Electron Beam Accelerator Facility (CEBAF) and the Relativistic Heavy Ion Collider (RHIC), as well as the NSF-supported National Superconducting Cyclotron Laboratory (NSCL), were developed in response to the long-range planning process. The first appearance of the need for a rare-isotope beam facility came from a community process that preceded the development of the 1989 LRP. In a series of NSAC activities that concept evolved into the DOE-supported Facility for Rare-Isotope Beams (FRIB). Similarly, the support of the nuclear physics community for fundamental nuclear science, focusing on an underground laboratory, stems from community-driven action expressed at the Astrophysics, Neutrinos, and Symmetries Town Meeting that took place before the 2002 LRP and that was highlighted in the white paper that emerged from that town meeting.

The LRP process has evolved into a fairly standard sequence of stages. It commences with a charge from DOE and NSF, usually involving one or more budget scenarios. Community involvement ensues, first through informal workshops and then through large town meetings representing the different subfields of nuclear science, as well as through meetings that focus on applications and on education and outreach. The town meetings engage a very large fraction of the active nuclear scientists in the United States. Each town meeting develops an extensive and detailed white paper that discusses the scientific and technical opportunities in each field and sets forth priorities and recommendations. The white papers serve as a valuable technical resource during the writing of the LRP itself.

The defining event in the LRP process is a week-long meeting of scientists from all areas and constituencies of the field, with agency representatives as observers and international invitees. The output, as noted above, is a set of recommendations and a fleshed-out discussion of the field, including basic research, facilities, applications, education and training, and national workforce needs, all in the context of and under the constraints of one or more budget scenarios.

This process has been extraordinarily successful. Every major new facility built for nuclear physics research in the last 30 years emerged from within the community and was carefully vetted by the broader nuclear science community via this process and has usually been accorded the highest priority for new construction, and for completion of construction, in more than one successive LRP. Conversely, no major facility has been built without strong endorsement in one or more LRP documents. On the one hand, this history implies a rather long birthing process for such initiatives; on the other, it allows careful weighing and ordering of priorities and lays the groundwork for the unanimous (or nearly unanimous) recommendations that emerge. This unanimity among scientists with widely different interests in the overall field is one of the most remarkable achievements of the LRP process.

The sequence of LRP documents has been enormously influential, both domestically (see above) and abroad. The process takes into account developments and initiatives occurring in other countries and, in turn, influences those initiatives considerably, as illustrated by the recent report of the Nuclear Physics European Collaborative Committee (NuPECC), *Perspectives of Nuclear Physics in Europe.*[2] The priorities in nuclear science laid out in the LRPs and in Europe, Japan, China, and elsewhere exhibit a very high degree of congruity, reflecting a clear worldwide consensus, driven by the physics questions at stake and the technical innovations that allow for their resolution, and a set of complementary approaches to answering these questions.

The committee strongly supports continued reliance on the nuclear science long-range plan process. The time and effort required by that process to develop consensus in the nuclear physics community has been particularly worthwhile in the construction of large facilities. However, as discussed below, the committee recognizes that excellent smaller and/or international opportunities must sometimes be seized before the next LRP process begins in order to ensure the maximum scientific impact.

## Planning in a Global Context

The way planning for new directions in nuclear science is done has evolved over the past several decades: first, as the field has become increasingly international, with large, complex projects pursued by teams of scientists from many nations, and, second, as countries in Asia and Europe have made, and continue to make, major investments in facilities and infrastructure for nuclear science. Highly successful examples of advances and of projects in nuclear science involving international partners are abundant in this report. Beginning in this millennium and extending to the end of the present decade, the worldwide investment in new facility construction will be well over $5 billion, with over 80 percent of this occurring outside the United States. Thus, planning is now done in a global context, taking into account the opportunities that lie outside the United States. That said, it is worth emphasizing that (as concluded by the OECD Global Science Forum Working

---

[2] European Science Foundation, 2010, Perspectives of Nuclear Physics in Europe, NuPECC Long Range Plan 2010. Available at http://www.nupecc.org/lrp2010/Documents/lrp2010_final_hires.pdf.

Group on Nuclear Physics in 2008[3]) the worldwide nuclear physics effort is, and for the foreseeable future should continue to be, focused on regional facilities with international participants and contributions rather than on global facilities. The issue that must be confronted is how to plan U.S. nuclear science involvement in foreign projects without sacrificing U.S. intellectual leadership.

U.S. loss of its intellectual leadership in nuclear science would gravely compromise its ability to capitalize on future discoveries in this critical area of science. It would also negatively impact the U.S. economy and safety as the country would not benefit from new technological developments in the field and would lose workforce trained in nuclear techniques. For the United States to maintain intellectual leadership in the field, it must develop and operate state-of-the-art facilities that yield scientific breakthroughs and open new frontiers of knowledge. This is what attracts leaders in the field to the United States and keeps them here and results in a new generation of leaders, some emerging in the United States and others being attracted here. At the same time, one of the best ways for us to build and maintain intellectual leadership is for U.S. scientists to do experiments at facilities overseas, as long as they are positioned and supported such that the scientific impact of their contributions is significant, and seen to be so, within a context of shared responsibility and shared achievement. Such an environment makes it possible to optimize scientific endeavors in a globalized world and also underlines how important it is for the United States to be a reliable and committed partner in such endeavors.

There is no one prescription for an optimal balance between the use of home and foreign facilities. In the LRP process, these questions are weighed, debated, and argued in an open and healthy way as people advocating different possible paths forward for U.S. nuclear science make their cases and critique other cases. Having international invitees present is important for the inside information and perspective that they bring. What is even more important, however, is the constructive push and pull between members of the U.S. community, who never hesitate to argue that the best way for the United States to build its leadership in a certain area of the field can be to invest in U.S. participation at overseas facilities if they see such a path as the best way to have a scientific impact, given that not everything can be done within the United States. Ultimately the U.S. program must fit within funding constraints that are determined by the federal government. Planning within these constraints, the nuclear science community attempts to optimize the program to maintain intellectual leadership.

---

[3] Organisation for Economic Co-operation and Development (OECD), 2008, Report of the Working Group on Nuclear Physics, May. Available athttp://www.oecd.org/dataoecd/35/41/40638321.pdf. Last accessed on August 26, 2011.

### The Need for Nimbleness

Until now, the focus has largely been on how the U.S. nuclear science community and the support agencies make decisions about large facilities. As discussed in the preceding section, LPRs involve a rather long birthing process, which is desirable in the case of large facilities since it allows for careful prioritization and for the emergence of near-unanimous support for major national initiatives. The committee recommends, however, that the U.S. planning processes need to become more nimble, streamlined, and flexible when it comes to initiating and managing smaller projects or seizing international opportunities. What is needed is not so much a change in the LRP process as much as (1) consideration of how the sophisticated new tools and protocols that have been developed for the successful management of the largest projects in nuclear physics can best be applied to projects at the other end of the size and cost scales and (2) recognition that sometimes excellent smaller-scale and/or international opportunities must be seized before the next LRP process begins in order to ensure that the scientific impact of U.S. participation is significant.

It is worth observing that the strategic management of science is carried out in quite different ways by different national governments. A comparison of the Japanese and U.S. systems is of interest. Japan has been extraordinarily successful in neutrino physics, among other areas. At the risk of oversimplifying, the Japanese system retains flexibility through a hierarchical structure and close communications between Cabinet members and senior science administrators. Decisions are made at a high level following careful but not unlimited review. This approach makes for a certain nimbleness that is less evident in the United States, where it sometimes appears that Congress, the funding agencies, and the researchers have become adversaries. Agencies answerable to Congress tend to become highly risk-averse and develop continuous review processes that, while effective in improving cost and schedule predictability, are themselves resource-intensive. Project-cost thresholds have been established at the funding agencies for implementing increasingly stringent levels of control and review, based on past experience. It can be tempting, however, to implement each control model well below its established cost threshold in the belief that the outcome can only be improved. It is important to keep in sight the goal of scientific research is achievement and discovery, and that cost and schedule predictability are tools to aid that process, not the primary objective.

Nimbleness is essential if the United States is to be an innovator and is to remain among the leaders in nuclear physics. Streamlined and flexible procedures for initiating and managing smaller projects are recommended. Among the changes that can be considered are the recognition that surprise, reassessment, and course-correction are a natural part of the research enterprise, that review and reporting requirements have reached the point of counterproductivity, and that discretion to move funding across "firewalls" within a discipline should rest with midlevel agency administrators under advisement from community leaders.

## A NUCLEAR WORKFORCE FOR THE TWENTY-FIRST CENTURY

### Challenges and Critical Shortages

The twenty-first century has brought a growing realization that a well-trained nuclear workforce of adequate size is crucial to address the serious challenges facing our world. The tragic suicide attacks of September 11, 2001, opened whole new classes of threat scenarios that require novel and sophisticated responses. There are significant new requirements and challenges in the fields of nuclear forensics, border protection, and nonproliferation, some of which are discussed in Chapter 3. And, present-day nuclear needs go well beyond those for threat reduction. The demand for a nuclear workforce for medicine, health physics, and energy is certainly not decreasing. All of these areas are important for national and world security and prosperity, yet their increasing needs come at a time when the nuclear workforce is shrinking.[4]

The workforce problem is one that has been building over several decades, but the severity of the situation has become obvious in the last decade. The fields of nuclear chemistry and radiochemistry have seen a steady decline in the number of new Ph.D.s from about 30 per year in the 1970s to fewer than 5 per year today.[5] The threat of a terrorist nuclear attack has brought the situation to a crisis level, as described, for example, in the 2008 report from the American Physical Society (APS) Panel on Public Affairs *Readiness of the U.S. Nuclear Workforce for 21st Century Challenges*[6] and the 2010 NRC report *Nuclear Forensics: A Capability at Risk.* Unless education programs are reinvigorated, the United States will lack the expertise to pursue the research needed to advance and maintain those fields of nuclear physics and radiochemistry that are intrinsic components of nuclear medicine for diagnosis and treatment, the handling and storage of reactor waste, detection of the trafficking of nuclear material, nuclear forensics, nuclear weapons and stockpile stewardship, and the development of new accelerator technologies.

The workforce problem is not limited to low-energy nuclear physics and radiochemistry; it affects all areas of applied nuclear science. The national laboratories

---

[4] American Physical Society, 2008, "Readiness of the U.S. Nuclear Workforce for 21st Century Challenges." Available online at http://www.aps.org/policy/reports/popa-reports/upload/Nuclear-Readiness-Report-FINAL-2.pdf. Last accessed on October 29, 2012.

[5] Ibid. and references therein.

[6] Ibid.

funded under DOE's National Nuclear Security Administration (NNSA) are charged with addressing nuclear science as it relates to security, both stockpile stewardship and threat reduction. A significant fraction of the workforce involved in this research was originally attracted to the field through collaborative basic nuclear physics research programs between the universities and the national labs. Recruits have come from all subfields of nuclear physics, and there is a strong track record of successfully applying nuclear physics techniques to national security needs. Some of the recent and more novel success stories are highlighted in Chapter 3. However, as in the case for radiochemistry, there is an increasing decline in the percentage of physics Ph.D.s graduating with expertise in nuclear physics at a time when workforce demands are growing.[7] The situation is most alarming when viewed in terms of the top research universities in the United States, where nuclear physics is now a major discipline in only a handful of physics departments.

One important positive factor that has worked against this decline is the broadening of the field. Over recent decades this broadening has taken place in the directions of fundamental symmetries, neutrino physics, cold atoms, open quantum systems, lattice quantum field theory, quark-gluon plasma, high-performance computing, and new areas of nuclear astrophysics—all of which are featured in earlier chapters. Continued expansion of the basic science domain in which nuclear physics and nuclear physicists play a leading role, driven by new discoveries, new opportunities, and new people, will further ameliorate the decline as new talent is drawn into this wider field. Even with this diversification of nuclear research, however, the workforce shortage will become acute unless a coordinated and integrated plan is implemented to build and sustain an appropriately sized workforce, coupled with the necessary research facilities. Some elements that could play an important role in such a plan are described below, but first the role of graduate students and postdocs as well as a crucial balance issue that must be at the forefront in any such planning are discussed.

### The Role of Graduate Students and Postdocs

Nuclear physics has a long tradition of preparing leaders for a wide range of applications of nuclear physics and its techniques, as well as future leaders in basic nuclear science research. Each one of these future leaders starts out as a graduate student, and most are also postdoctoral scholars. Scientists in these early stages of their careers play critical roles in all aspects of the basic research enterprise, while preparing to become leaders in basic and applied research and development.

---

[7] American Physical Society, 2008, "Readiness of the U.S. Nuclear Workforce for 21st Century Challenges." Available online athttp://www.aps.org/policy/reports/popa-reports/upload/Nuclear-Readiness-Report-FINAL-2.pdf. Last accessed on October 29, 2012.

Graduate students in nuclear science are expected to play key roles in all aspects of the research program to which their advisors introduce them. They learn how to develop new experimental tools or theoretical techniques, mount experiments or perform calculations that may require massive computations, analyze and interpret data and results, disseminate their results in written and oral form to a variety of audiences, and ultimately learn how to propose new experiments or initiate new theoretical developments. Graduate students also often join their more senior scientist advisors in mentoring undergraduate students in summer research projects. Postdoctoral scholars are taking their first steps as independent nuclear scientists in their own right. They are developing their own research agendas, and collaborating with and helping to mentor graduate students. Postdocs have mastered the tools and skills needed to advance their research in nuclear physics, and they have the time to devote their full energies to doing so. Very often in a collaboration it is the postdocs who play the most important role in pushing a project to completion and making new discoveries. Together, graduate students and postdoctoral fellows are the engines of research in nuclear science.

The 2004 report of the Nuclear Science Advisory Committee (NSAC), *Education in Nuclear Science,* includes a survey of Ph.D. recipients in nuclear science 5-10 years after their doctorates.[8] It found that 40 percent of them work outside universities, colleges, and national laboratories. They are contributing to the nation's needs for nuclear scientists, addressing the nation's challenges in security, health, energy, and education as well as contributing innovations in technology and business that help to drive our economy. The number of new Ph.D. recipients in all of nuclear science is about 80 per year and has been relatively constant over the past decade. Today nuclear science accounts for only about 5 percent of all the Ph.D. recipients in physics and astronomy, a percentage that has been declining in recent years as the number of Ph.D. recipients in physics in the United States has been increasing. This decline is often due to an insufficient number of research assistantships or of faculty mentors. Given the many ways that nuclear scientists contribute to the needs of the United States and the challenges and critical shortages that are described above, and given the wealth of basic research opportunities described throughout this report, there is a demonstrated need to increase the number of graduate students in nuclear science. The committee therefore endorses the 2004 NSAC report's recommendation for an increase of 20 percent in the number of Ph.D. students in nuclear science over the next 5 to 10 years.

---

[8] Nuclear Science Advisory Committee (NSAC), 2004, *Education in Nuclear Science.* Available at http://science.energy.gov/~/media/np/nsac/pdf/docs/nsac_cr_education_report_final.pdf. Last accessed on October 29, 2012.

## Balance of Investments in Facilities and Universities

Advancing the research frontier in many areas of science has required the construction of large research facilities at national laboratories or universities, to serve a large base of users from many universities and colleges. This has been true for a while in particle physics, is now true in nuclear physics and astronomy, and is becoming more common in areas like condensed matter physics, biology, and chemistry. In this circumstance, strong partnerships between universities and colleges and the institutions where the facilities are located are essential to reaping scientific rewards from the significant national investments that it takes to construct and operate world-class facilities. The relationship between the facilities and the college and university groups that use them is symbiotic: Without the facilities, the science cannot be done; without the colleges and universities, there would ultimately be no people doing the science. Universities are where scientific advances attract the brightest young minds into nuclear science and where future nuclear scientists make their first research contributions. Strong partnerships between the national laboratories and universities that host major facilities and universities and colleges around the country are essential to ensure that there is a next generation of nuclear scientists. Even as the pressure for more resources to support operations of the major facilities themselves becomes more acute, the long-term health of the field of nuclear science and of the nuclear science workforce that is needed by the nation requires that a balance be established and maintained between the needs of university and college programs on the one hand, and major facilities and national laboratories on the other. To this end, funding for educating, training, mentoring, and supporting the research of budding nuclear scientists from undergraduates through junior faculty is an essential component of moving the science forward. This component of the national effort is as necessary as the facilities to advance nuclear science research and to provide the nation with the nuclear workforce, expertise, and applications that it critically needs.

## Mechanisms for Ensuring a Robust Pipeline

The agencies that support nuclear science research should be creative in finding new ways in which to partner with university and college groups in order to ensure a robust pipeline. Several mechanisms for doing so are listed, some of which are already in operation. These suggestions should not be interpreted as exclusive. The symbiotic relationship between educational institutions and facilities needs strengthening across a variety of fronts. There is no one solution; what is provided below is a selection of actions that are contributing to or would work toward the desired goals. Formal recommendations of the committee that include several of these steps are set forth in Chapter 6.

- The introduction of a national fellowship program for graduate students in nuclear science or entering the field would identify and support the best students in the nation pursuing nuclear science research, with the goal of attracting the highest caliber students into the field. While the fellowships would be awarded to students at U.S. universities, the research supervisor could be a national laboratory scientist. Care should be taken in the selection process so that it is receptive to exceptionally good students located anywhere, as one benefit should be to identify and recognize outstanding students wherever they may be found. A national prize fellowship would promote nuclear physics to U.S. undergraduate students and first-year graduate students considering such opportunities. Models for the nuclear physics fellowship program can be found in other fields; examples within the DOE include its fellowships in the computational sciences, the fusion energy sciences, and the stewardship sciences.

- The introduction of a national prize fellowship program for postdoctoral researchers in nuclear science would provide similar encouragement and support for those at the beginning of their research careers. Winning a prestigious named fellowship in a national competition will raise the profile of a researcher at an early stage and enhance the visibility of the brightest early-career nuclear scientists in the academic world. Giving the winners both support and freedom as they launch their research careers will maximize the scientific impact of these future leaders of the field at a time when their abilities are fully developed and their energies are devoted solely to research. Furthermore, the existence and visibility of such a program will serve to attract highly talented students to do graduate work in nuclear science. The fellowship procedures should be designed to achieve a balance among giving winners freedom to choose how they want to develop their careers, ensuring that they find a host university or national laboratory that welcomes and mentors them, and ensuring that over time many institutions get to host winners working in many subfields. The astronomy community has developed mechanisms for achieving these goals for Hubble and Einstein postdoctoral fellows. Funding agencies could also support an annual symposium, for example, coordinated with the Division of Nuclear Physics (DNP) of the APS fall meeting, at which current holders of the fellowship give talks. Such a symposium, patterned on NASA's successful Hubble Fellows Symposia, whose webcasts are often viewed by faculty search committees, would draw further attention to the fellowship and to nuclear science while building a sense of community among emerging leaders working in diverse areas of the science.[9]

- Future graduate students in nuclear science are undergraduate students first. The participation of nuclear scientists as research mentors for undergraduate students during the summer—for example, in NSF-funded research experience for undergraduates (REU) programs at universities as well as in similar programs in national laboratories and at other universities—plays an important role in attracting some of these students into nuclear science and in adding nuclear physics to the educational experience of physics students who go on to careers across the full spectrum of science and technology, from medical doctors to engineers to teachers to scientists in other fields. Each year, the Conference Experience for Undergraduates (CEU) program brings over 100 undergraduates (typically about one third are women) with nuclear physics research experience to the annual conference of the DNP (see Box 5.1). The CEU program gives undergraduates the opportunity to present their research to the nuclear science community, provides them with a capstone experience for their research projects, builds a sense of community and collective accomplishment among future nuclear scientists at a very early stage in their careers, and exposes them to the full breadth of the field. The program has met with tremendous success, the enthusiasm of the DNP community, and support from NSF and DOE as well. The CEU poster session is one of the best attended events at the annual meeting of the DNP.

- The NSF is to be commended for its CAREER awards and the DOE for its Early Career Awards and its former Outstanding Junior Investigator program. In addition to providing their recipients with support at a critical career stage, these awards help to highlight their outstanding achievements and increase their recognition at their home institutions and throughout the community.

- It is essential to keep nuclear science groups at universities, colleges and national laboratories vital and vigorous. This requires that retiring or otherwise departing nuclear scientists be replaced and, where existing efforts are below optimal size, new faculty or staff positions be created. Bridging support is a very good investment when it facilitates the creation of new faculty and staff positions.

- Competitive awards for shared research instrumentation awards, such as the Major Research Instrumentation program of the NSF, work well in nuclear physics because the funding is matched reasonably well to typical

---

[9] Portions of this paragraph were adapted from NSAC Subcommittee on Nuclear Theory, 2003, *A Vision for Nuclear Theory*. Available at http://www.nucleartheory.net/docs/NSAC_Report_Final8.pdf.

**Box 5.1**
**Conference Experience for Undergraduates:**
**A Capstone Experience with an International Reach**

Held annually since the fall of 1998, the CEU was developed and organized to provide undergraduate students who had conducted nuclear science research the opportunity to attend the fall meeting of the Division of Nuclear Physics (DNP) of the APS, and to present their research to the professional nuclear science community. The program is supported by the NSF, DOE (at six national accelerator laboratories), and the DNP. The number of participating students has grown to well over 100 annually, a one third of whom are women. Travel grants are awarded on a competitive basis, and all participating students receive lodging. Activities for the students, in addition to participation in regular conference activities, include the research poster session, two nuclear physics seminars targeted to the advanced undergraduate level, an ice cream social, and a graduate school fair. The program has met with tremendous success and is enthusiastically supported by the DNP community, with the CEU poster session one of the best-attended events at the conference. In the fall of 2009, at the joint meeting of the APS and the Japanese Physical Society, held on the Big Island of Hawaii, 116 American and 20 Japanese undergraduate students participated in the CEU (Figure 5.1.1). Two previous joint APS/JPS meetings (fall 2001 and 2005 on the island of Maui) also provided an opportunity for undergraduate students from the two countries the rare opportunity to meet, interact, and initiate long-term connections.

FIGURE 5.1.1 Many of the more than 100 undergraduate students who conducted research in nuclear physics and then presented the results from their research at the 2009 meeting of the APS Division of Nuclear Physics in Waikoloa, Hawaii. SOURCE: Courtesy of W. Rogers, Westmont College.

project sizes. With a competition every year and resources given up front for successful proposals, this program stimulates the nimbleness described above that is so critical for medium-sized projects. Similarly, modest grants to universities for infrastructure improvements or equipment purchases can often be cost-shared with university resources. Such funding enables university groups to train students to build state-of-the-art experimental equipment at their home institutions. More of these smaller grants should be made available, as they can help to reverse the erosion of the infrastructure for nuclear physics at universities. They are also perfect examples of the symbiosis between universities and facilities: Building detectors or detector components at a college or university draws students into the field, and both the apparatus and the students subsequently play crucial roles in doing nuclear science at the large facilities.

- The DOE has recently funded several Topical Collaborations in Nuclear Theory. While it is too soon to fully evaluate the scientific impact that results, these efforts appear to be strengthening connections between groups at national laboratories and universities, creating new bridge-funded faculty positions, and invigorating small university groups, thereby advancing the goals described above. There were many more strong proposals for such collaborations than could be funded in the initial round; indeed the call for proposals had the effect of stimulating nuclear theorists to come together in new ways, developing new proposals and new ideas. A new competition for future such collaborations would build on these successes and this momentum.

- As noted above, the successful broadening of the field of nuclear physics in recent decades into many of the areas described earlier in this report has been crucial to maintaining a nuclear physics presence in universities and colleges and a nuclear workforce for national needs and has played a key role in keeping the field intellectually vibrant. Further broadening is occurring—for example, when experimental nuclear physicists employ their expertise to advance the search for dark matter and theoretical nuclear physicists take on the challenges of cold atom experiments—and will continue to occur in the future. As new discoveries, new opportunities, and new people expand the domain in which nuclear science and nuclear scientists play a leading role, the funding agencies should help nurture such expansion.

## Broadening the Nuclear Workforce

The population of the United States has become increasingly diverse, and individuals of gender, race, and ethnicity who are underrepresented in the physical sciences, including women and racial and ethnic minorities, now constitute a large majority of the student population in colleges and universities nationwide. The nuclear science community has developed a number of strategies to recruit undergraduates into nuclear science proactively, often with particular attention to the underrepresented groups. As described above, mechanisms include the many REU programs at universities and colleges, similar programs at the national laboratories, and the CEU. Nuclear scientists at colleges and universities also teach undergraduate science and engineering students, reach out to minority-serving institutions and their faculty, and develop bridge programs that foster the transition of students from smaller colleges to graduate studies at research universities. This engagement adds nuclear science to the educational experience of students who go on to careers across the full spectrum of science and technology. With revitalized undergraduate and master's programs in nuclear engineering, there are new opportunities for nuclear scientists to partner with their colleagues to enhance nuclear science education, to reach out to students from traditionally underrepresented backgrounds, and to prepare students for a broad range of opportunities in nuclear science, engineering, and technology. Many institutions also reach out to schools and to their communities, visiting schools and opening the doors of their laboratories to school teachers and to students, both informally and via varied, more formal programs, including some that give them the opportunity to participate in nuclear research. All of these efforts, each in its own way, serve to broaden the nuclear workforce. Support for such programs from the funding agencies is a way to recognize the need for a robust pipeline of education and training for the next generation of nuclear scientists.

## Highlight: Nuclear Crime Scene Forensics

An attack using a nuclear device on one of our cities would be both catastrophic and world changing. President Obama describes it as the single biggest threat to U.S. security. In the event of such an attack, a set of urgent and crucial questions would have to be answered: What was exploded? Who was responsible? Do they have more? Was the device improvised or sophisticated? Did they steal it or have help making it? Is the material reactor-grade or weapons-grade fuel? How old is it? Detonation of a radiological device, a "dirty bomb," could also result in widespread contamination and public concerns. Nuclear forensics is the technical means and set of scientific capabilities that, in the event of such attacks, would be used to answer these questions (see Figure FOR 1).

Nuclear forensics involves the analysis and evaluation of postdetonation debris following a nuclear explosion. It is also essential in the analysis of unexploded devices or material that have been seized. The basic idea behind nuclear forensics is the same as that behind the

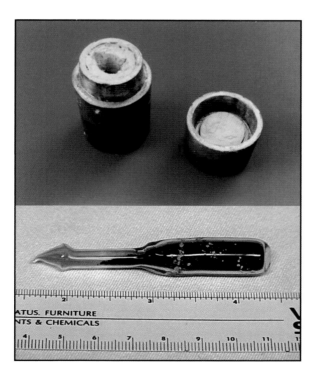

FIGURE FOR 1 Lead container and the glass ampoule containing highly enriched 235-uranium seized in Bulgaria in 1999 and analyzed at Lawrence Livermore National Laboratory for forensic purposes. SOURCE: "Forensic Analysis of a Smuggled HEU Sample Interdicted in Bulgaria," UCRL-ID-145216, August 2001.

*continued*

analyses of stellar nucleosynthesis. In stellar astrophysics, the debris from nuclear reactions inside a star is used to infer detailed information about the nature of the star and the reactions, including the mass, density, compositional layers, and temperature of the star. Very similar analyses are used in nuclear forensics to answer the pertinent questions for an exploded nuclear device here on Earth.

During the nuclear testing era, the United States and other countries used radiochemistry techniques to characterize the explosions from collected debris. The application of the radiochemistry experience and techniques developed over those decades has proved invaluable in generating the key concepts underlying nuclear forensics. The current forensic capabilities have been used in postdetonation exercises in which the national laboratories have demonstrated that they can characterize nuclear debris and other forensic data and can infer the key design features for a variety of hypothesized nuclear explosive devices.

Several key nuclear physics concepts apply in forensics. One is the use of the natural radioactive decay of a nuclear material to determine its age. Plutonium, which does not occur naturally on Earth, would originally have been produced in a reactor with some unavoidable plutonium-240 and plutonium-241 content. The decay of plutonium-241 into americium-241 with a 14.4-year half-life indicates the time lapsed since production. This is analogous to the way in which carbon dating is used to determine the age of some material. Key information regarding the origin of plutonium material can also be obtained from the ratio of the different plutonium isotopes—for example, whether it is weapons grade or reactor grade and the total reactor neutron fluence (or burnup) to which it was originally exposed.

The design of a detonated device from the explosion debris can be inferred from the shape of the neutron flux spectrum, which serves as a fingerprint for the design. An analogous fingerprint for nuclear reactors is the average energy of the neutron spectrum, from which one can deduce whether the reactor is a thermal light or heavy water reactor or a fast reactor, which will point to whoever designed the reactor. To extract the shape of the neutron flux spectrum for a detonated nuclear device, forensics takes advantage of the very different energy dependencies of nuclear cross sections. Useful nuclear reactions fall into three categories: (1) neutron capture, which characterizes the low-energy part of the spectrum, (2) inelastic neutron scattering to nuclear isomers, which characterizes the fission neutron component of the spectrum, and (3) threshold *(n, 2n)* reactions, which characterize the high-energy component of the spectrum from fusion neutrons.

As an example, let us consider how we could use the americium-241 present in the fuel of a plutonium-based nuclear device to extract information about the shape of the neutron spectrum. During the explosion, some of the americium-241 will be transmuted to americium-240 through the *(n,2n)* reaction; this americium-240 could not have been present before detonation because it lives for only 2.1 days. Producing americium-240 from americium-241 requires the presence of neutrons with energies of at least 6.67 MeV, because the nuclear reaction involved is a so-called threshold reaction. Thus, the relative abundance of americium-240 provides unique information about the high-energy component of the neutron spectrum (see Figure FOR 2).

Detailed studies of the implication of the production of americium-240 or other isotopes of interest require that all of the significant paths for producing and destroying these isotopes during a nuclear explosion be known. To address this requirement, a number of multiinstitutional programs are performing a series of measurements. One of the collaborations is attempting the first measurement of the fission of americium-240, a very challenging undertaking for several reasons. The first two challenges are the production of a significant quantity of americium-240 and its subsequent chemical separation to form a target. These are being carried out at LBNL using the 88-in. cyclotron and radiochemistry facilities. Stewardship Science fellow Paul Ellison, who is featured in the Highlight "Future Leaders" between Chapters 5 and 6, is part of this team. Only very small targets of americium-240 will be possible, making the fission cross section measurement another challenge. For the fission measurement, the very high neutron intensity capability of the Lead Slowing-Down Spectrometer (LSDS) at the Los Alamos Neutron Science Center will be used. The LSDS will also be used to measure another fission reaction of interest for uranium devices—in particular, the fission of uranium-237. Like americium-240, uranium-237 is very radioactive and difficult to produce and chemically separate. In these experiments, uranium-237 is produced by irradiation of uranium-236 in the high-flux reactor at Oak Ridge National Laboratory. Uranium-237 has a half-life of 6.75 days, and handling the irradiated sample to chemically separate uranium-237 requires use of Oak Ridge's hot cell facilities. Once the targets are fabricated, the fission measurements will be carried out at Los Alamos using the LSDS.

The committee hopes that the validity of our nation's nuclear forensics schemes will never need to be tested directly. Confidence in the program capability is greatly enhanced by the tight coupling between the fundamental nuclear data community, radiochemists, and the weapons design community. It hopes as well that broadcasting the capabilities of nuclear forensics to identify the source of a nuclear device and its fuel will deter advocates of such unthinkable acts.

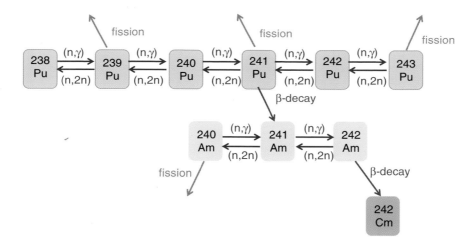

FIGURE FOR 2 Plutonium (Pu) and americium (Am) reaction chains. One of the pieces in a nuclear forensics puzzle is the amount of plutonium-241 in a nuclear device, which can be determined after an explosion. The figure illustrates all of the neutron-induced reactions that need to be understood to deduce the original amount of plutonium-241 in the device and the fluence of high-energy neutrons in the subsequent explosion. These include reactions on americium-241, the daughter from radioactive decay of plutonium-241. The relative abundance of americium-240 in the explosion debris would provide key information for forensic analyses, because americium-240 can only be produced by high-energy ($E_n$ > 6.67 MeV) neutrons. SOURCE: Courtesy of A. Haynes, Los Alamos National Laboratory.

# 6

# Recommendations

Recent strides have brought nuclear science to the threshold of major advances in understanding the atomic nucleus and the possible applications of nuclear techniques. This progress is largely due to technological breakthroughs that allow for a deeper understanding of nuclei, their constituents, and the role of nuclei in the cosmos and to the development of many practical applications.

To remain vital, nuclear science must attract and retain top talent by providing a dynamic environment where innovation can flourish. This is possible only with the strong support of research and a diverse portfolio of wisely selected and efficiently operated facilities needed to carry out the program. Nuclear science research is diverse in scale, involving small and large groups of scientists and students from universities and research laboratories. The integration of research and teaching at universities is the natural environment for the incubation of new ideas and the education and training of the scientific workforce. Small university-based facilities support important research in nuclear structure, nuclear astrophysics, and fundamental symmetries. The infrastructure at such universities—a local accelerator, the technologies for advanced detectors, and instrumentation—enables university scientists and students to lead and to make important contributions to significant initiatives in nuclear science. Support for smaller-scale operations at the universities along with the operations of the major user facilities are indispensable components of the U.S. nuclear science program. In recent years the resources that go to the major facilities have made up an increasing fraction of the total budget, especially that devoted to the Department of Energy (DOE) portion. Continuing to increase the share reserved for facilities operations at the expense of the research budget is not sustainable.

Recently the DOE nuclear physics program took over the stewardship of the National Isotopes Program. This program continues to grow in importance as the uses of nuclear techniques in medical imaging and therapeutic procedures accelerate. The isotopes program and nuclear physics have been ideal partners and the impact has been positive on both sides. The overall costs of the isotope program have not impacted the budget in other areas of nuclear physics, but the synergy between the two programs has resulted in more efficient operations on the isotope production side and more opportunities for research and development on the nuclear physics side, including opportunities for accelerator physicists to develop new techniques in isotope production. Another example of mutual benefit is the inclusion of the isotopes program in the Small Businesses Innovation Research program and the Early Career and Graduate Fellowship program in nuclear physics, while isotopes produced for research do not incur additional costs for nuclear physics. The continued health of the isotopes program, in particular, reinforces the need for a workforce trained in nuclear science and the need for strong university programs to provide that training.

## FOLLOWING THROUGH WITH THE LONG-RANGE PLAN

The Department of Energy Office of Nuclear Physics operates three user facilities for nuclear physics research in the United States: the Argonne Tandem Linear Accelerator System (ATLAS), the Continuous Electron Beam Accelerator Facility (CEBAF) at the Thomas Jefferson National Accelerator Facility (JLAB), and the Relativistic Heavy Ion Collider (RHIC) at Brookhaven National Laboratory (BNL). The National Science Foundation (NSF) operates one nuclear physics user facility, the National Superconducting Cyclotron Laboratory (NSCL) at Michigan State University (MSU). The Spallation Neutron Source (operated by the Department of Energy Office of Basic Energy Research) hosts the Fundamental Neutron Physics Beam line at Oak Ridge, which provides cold neutron beams for nuclear physics research. A number of NSF- or DOE-supported smaller facilities at universities make unique contributions to the field. Two large centers—the Institute for Nuclear Theory and the Joint Institute for Nuclear Astrophysics—serve the nuclear physics community and facilitate strong connections to other fields of science.

As a consequence of a systematic long-range planning process and strategic investments by funding agencies, nuclear physicists have access to U.S. facilities with world-class physics programs. CEBAF is undergoing major upgrades of its accelerator and detectors, opening up a new frontier in nuclear electroweak physics. The upgrade of the CEBAF electron accelerator and detector systems will enable the measurements required to search for exotic mesons, a fundamental prediction of quantum chromodynamics (QCD). It will also further our understanding of how quarks and gluons form nucleons and nuclei, the most fundamental building blocks of visible matter, and enable a precision test of the Standard Model. The luminosity at RHIC has been increased and the detectors have been upgraded, providing the opportunity to conduct experiments that can make more incisive measurements of the contributions of quarks and gluons to the spin of the proton and capitalizing on RHIC's recent discovery of a liquid phase for quarks and gluons. U.S.-sponsored scientists participate in experiments at the Large Hadron Collider (LHC) that should bring new ways of exploring this new phase of matter. Neutrino experiment discoveries at underground laboratories already required the first revision to the Standard Model in four decades, and new experiments to answer important questions about neutrinos are under way or planned in the United States and around the world. The low-energy nuclear physics user facilities, ATLAS and NSCL, are developing new paradigms of nuclear structure. These facilities, together with low-energy stable beam and neutron beam facilities, provide an array of beams and equipment required for the understanding of nuclear structure and nuclear reactions responsible for element production in stars and stellar explosions. These are also the tools needed for giving society innovative applications of nuclear science. The Facility for Rare Isotope Beams (FRIB), now under construction at MSU, will provide unique capabilities within an expanding worldwide arsenal of rare isotope facilities. The Fundamental Neutron Beam line is poised to begin its research program, which includes experiments with enormous discovery potential by making unprecedented tests of the fundamental symmetries of nature. Nuclear theorists, computer scientists, and applied mathematicians are taking advantage of state-of-the art supercomputers to carry out calculations of previously intractable complexity, leading to new understanding of nuclear structure and reaction dynamics, supernova explosions, nucleon structure, and quark-gluon plasma properties. With all these new tools, the U.S. nuclear physics community is poised to make important discoveries in the coming decade.

**Finding: By capitalizing on strategic investments, including the ongoing upgrade of the continuous electron beam accelerator facility (CEBAF) at the Thomas Jefferson National Accelerator Facility and the recently completed upgrade of the relativistic heavy ion collider (RHIC) at Brookhaven National Laboratory, as well as other upgrades to the research infrastructure, nuclear physicists will confront new opportunities to make fundamental discoveries and lay the groundwork for new applications.**

**Conclusion: Exploiting strategic investments should be an essential component of the U.S. nuclear science program in the coming decade.**

As an outcome of careful long-range planning by the community and funding agencies, two large construction projects for nuclear science research facilities are now on their way in the United States. For the coming decades, these projects will enable the U.S. nuclear science community to continue to make discoveries that push the boundaries of knowledge and benefit society. One of the projects is the upgrade to CEBAF. The other is a new facility, FRIB, which will be the world's most powerful device for studying the properties of exotic unstable atomic nuclei and their reactions, providing unique research opportunities for the United States and the international community. Data to date on exotic nuclei are already beginning to revolutionize our understanding of the structure of atomic nuclei. FRIB will enable experiments in uncharted territory at the limits of nuclear stability. FRIB will provide new isotopes for research related to societal applications as it addresses long-standing questions about the astrophysical origin of the elements and the fundamental symmetries of nature.

**Finding: The Facility for Rare Isotope Beams is a major new strategic investment in nuclear science. It will have unique capabilities and will offer opportunities to answer fundamental questions about the inner workings of the atomic nucleus, the formation of the elements in our universe, and the evolution of the cosmos.**

**Recommendation: The Department of Energy's Office of Science, in conjunction with the state of Michigan and Michigan State University, should work toward the timely completion of the Facility for Rare Isotope Beams and the initiation of its physics program.**

Nuclear scientists have a long and impressive record of scientific contributions with low background experiments at underground sites. In the most recent decade this work contributed to revolutionary discoveries about the neutrino. Further crucial experiments, including delicate experiments attempting to discover neutrinoless double-beta decay, are being mounted. Nuclear scientists are participating in experimentation that could lead to the discovery of rare nuclear interactions. This work could disclose the origin of dark matter, which may account for five times as much mass as that of all identified matter in the universe. An underground nuclear physics accelerator is being designed to study the nuclear reactions that are important to astrophysical processes associated with late stellar evolution. The nuclear physics community has participated in the exercise to make the scientific case for a deep underground laboratory in the United States. With private, state, and DOE funding, the Sanford Underground Research Facility (SURF), at moderate depth, will provide a home for experiments in dark matter and neutrinoless double-beta decay at the Homestake site in South Dakota. The scientific case for a deep underground laboratory in the United States remains compelling.

> **Recommendation: The Department of Energy, the National Science Foundation, and, where appropriate, other funding agencies should develop and implement a targeted program of underground science, including important experiments on whether neutrinos differ from antineutrinos, on the nature of dark matter, and on nuclear reactions of astrophysical importance. Such a program would be substantially enabled by the realization of a deep underground laboratory in the United States.**

## BUILDING A FOUNDATION FOR THE FUTURE

Universities are crucibles in which new ideas in nuclear science emerge, scientific advances attract the brightest young minds to the field, and future nuclear scientists make their first research contributions. In nuclear science as well as more generally, American universities are unparalleled engines of scientific innovation. University laboratories and research programs play leading roles in advancing all the frontiers of nuclear science that have been described as well as in developing its applications.

> **Finding: The dual role of universities—education and research—is important in all aspects of nuclear physics, including the operation of small, medium, and large facilities, as well as the design and execution of large experiments at the national research laboratories. The vitality and sustainability of the U.S. nuclear physics program depend in an essential way on the intellectual environment and the workforce provided symbiotically by universities and the national laboratories. The fraction of the nuclear science budget reserved for facilities operations cannot continue to grow at the expense of the resources available to support research without serious damage to the overall nuclear science program.**

> **Conclusion: In order to ensure the long-term health of the field, it is critical to establish and maintain a balance between funding of operations at facilities and the needs of university-based programs.**

With a strong, broad university program and large-scale, versatile facilities, nuclear science will build upon its track record of discovery and innovation and will help to build the skilled workforce needed for a healthy economy and to meet the energy, medical, and national security challenges that are faced today. However, the symbiotic relationship between universities and facilities needs strengthening across a variety of fronts in order to ensure a robust pipeline. In Chapter 5 we provide a selection of actions that would work toward the desired goals. Included among them are steps to encourage those entering graduate school to consider nuclear physics as an area of study. Graduate students play a key role in all aspects of the research programs of their advisors and help to fill the nation's need for nuclear scientists, not only in basic research but in the many areas described in Chapter 3 to which nuclear physics contributes. Here the committee recommends one such step:

> **Recommendation: The Department of Energy and the National Science Foundation should create and fund two national competitions: one a fellowship program for graduate students that would help recruit the best among the next generation into nuclear science and the other a fellowship program for postdoctoral researchers to provide the best young nuclear scientists with support, independence, and visibility.**

The rapid escalation in the power of computers is having an impact in all areas of human activity. The coming generation of extreme-scale computing resources will be required to make desired breakthroughs in key areas of nuclear physics. Nuclear physicists, computer scientists, and applied mathematicians are presently taking advantage of state-of-the art supercomputers to carry out very complex calculations, leading to new understanding of, and predictive capabilities for, nuclear forces, nuclear structure and reaction dynamics, hadronic structure, matter under extreme conditions, stellar evolution and explosions, and accelerator science. It is essential for the future health of nuclear physics that the theoretical nuclear science community have a clear strategy for exploiting the rapidly increasing power of modern computing for the benefit of their science.

> **Recommendation: A plan should be developed within the theoretical community and enabled by the appropriate sponsors that permits forefront computing resources to be exploited by nuclear science researchers and establishes the infrastructure and collaborations needed to take advantage of exascale capabilities as they become available.**

Nuclear science is an international effort involving cooperation and competition between scientists from different institutions and different nations and utilizing a broad range of large- and smaller-scale research facilities. In order to maintain their scientific leadership and continue to come up with innovations, U.S. nuclear scientists must operate in an environment that enables them to seize new opportunities that might arise in the United States or abroad in a timely manner so that they are fully engaged in the discoveries that result. As discussed in more detail in the section "The Need for Nimbleness" in Chapter 5, streamlining the sponsoring agencies' procedures for initiating and managing projects, especially smaller-scale projects whose risks are more easily manageable and whose potential for discovery and/or applications is large, is essential for a program that can keep up with, and indeed lead, the global community.

> **Finding: The range of projects in nuclear physics is broad, and sophisticated new tools and protocols have been developed for successful management of the largest of them. At the other end of the scale, nimbleness is essential if the United States is to remain competitive and innovative on the rapidly expanding international nuclear physics scene.**

> **Recommendation: The sponsoring agencies should develop streamlined and flexible procedures that are tailored for initiating and managing smaller scale nuclear science projects.**

Without gluons, there would be no neutrons or protons and no atomic nuclei. Gluon properties in matter remain largely unexplored and mysterious. An electron ion collider facility would provide unprecedented capability for studies that are essential for understanding the fundamental structure of visible matter, including (1) precision imaging of quarks and gluons to determine the spin, flavor, and spatial structure of the nucleon and (2) definitive measurements of the gluon fields in nuclei in a regime in which they are expected to be both strong and universal.

> **Finding: An upgrade to an existing accelerator facility that enables the colliding of nuclei and electrons at forefront energies would be unique for studying new aspects of quantum chromodynamics. In particular, such an upgrade would yield new information on the role of gluons in protons and nuclei. An electron-ion collider is currently under scrutiny as a possible future facility.**

> **Recommendation: Investment in accelerator and detector research and development for an electron-ion collider should continue. The science opportunities and the requirements for such a facility should be carefully evaluated in the next Nuclear Science Long Range Plan.**

# Appendixes

# A

# Statement of Task

The new 2010 NRC decadal report will prepare an assessment and outlook for nuclear physics research in the United States in the international context. The first phase of the study will focus on developing a clear and compelling articulation of the scientific rationale and objectives of nuclear physics. This phase would build on the 2007 NSAC Long-range Plan Report, placing the near-term goals of that report in a broader national context.

The second phase will put the long-term priorities for the field (in terms of major facilities, research infrastructure, and scientific manpower) into a global context and develop a strategy that can serve as a framework for progress in U.S. nuclear physics through 2020 and beyond. It will discuss opportunities to optimize the partnership between major facilities and the universities in areas such as research productivity and the recruitment of young researchers. It will address the role of international collaboration in leveraging future U.S. investments in nuclear science. The strategy will address means to balance the various objectives of the field in a sustainable manner over the long term.

# B

# Meeting Agendas

**FIRST MEETING**
**WASHINGTON, D.C.**
**APRIL 9-10, 2010**

**Friday, April 9, 2010**

*Closed Session*
7:30 am
*Open Session*

| | | |
|---|---|---|
| 1:00 pm | Welcome and introductions | Stuart Freedman, Chair<br>Ani Aprahamian, Vice-Chair |
| 1:10 | Perspectives from the National Science Foundation (NSF) | Joseph Dehmer, NSF |
| 1:50 | Perspectives from the Department of Energy (DOE) | Tim Hallman, DOE |
| 2:30 | Perspectives on the DOE/NSF Long Range Plan | Bob Tribble, Texas A&M University |
| 3:10 | Break | |
| 3:20 | Perspectives from the last decadal survey | John Schiffer, Argonne National Laboratory |

| | | |
|---|---|---|
| 4:00 | Setting scientific priorities, developing science policies | J. Patrick Looney, Brookhaven National Laboratory |
| 4:45 | Open microphone discussion | |
| 5:30 | Reception | |

*Closed Session*

## Saturday, April 10, 2010

*Closed Session*

## SECOND MEETING
## WASHINGTON, D.C.
## JULY 12-14, 2010

## Monday, July 12, 2010

*Closed Session*
*Open Session*

| | | |
|---|---|---|
| 8:10 am | Welcome and introductions | Stuart Freedman, Chair Ani Aprahamian, Vice-Chair |
| 8:15 | Perspectives from Japan | Shoji Nagamiya, Japan Proton Accelerator Research Complex (by videoconference) |
| 9:15 | Perspectives on high-performance computing, nuclear reactors | Robert Rosner, University of Chicago |
| 10:15 | Break | |
| 10:30 | Perspectives on nuclear physics in Latin America | Ricardo Alarcon, Committee member |
| 11:30 | Open discussion on morning's presentations | |
| 11:45 | Lunch | |
| 12:45 pm | Perspectives from industry | Eckert & Ziegler Isotope Products, Inc. (by videoconference) |

| 1:45 | Perspectives on nuclear theory | Berndt Mueller, Duke University |
| 2:45 | Break | |
| 3:00 | Perspectives on international activities | Walter Henning, Argonne National Laboratory |
| 4:00 | Open discussion on afternoon's presentations | |

*Closed Session*

## Tuesday, July 13, 2010

*Closed Session*

## Wednesday, July 14, 2010

*Closed Session*
*Open Session*

| 10:00 am | Perspectives on computational needs | Steven Koonin, Department of Energy |

*Closed Session*

| 1:00 pm | Meeting adjourns | |

## THIRD MEETING
## NEWPORT BEACH, CALIFORNIA
## SEPTEMBER 22-23, 2010

## Wednesday, September 22, 2010

*Closed Session*
*Open Session*

| 8:40 am | Welcome and introductions | Stuart Freedman, Chair<br>Ani Aprahamian, Vice-Chair |
| 8:45 | Perspectives on nuclear astrophysics | Michael Wiescher, University of Notre Dame |
| 9:45 | Break | |

| 10:00 | Double-beta decay experiments | John Wilkerson, University of North Carolina |
|---|---|---|
| 11:00 | Electron-ion collider | Allen Caldwell, Max-Planck-Institut für Physik (by videoconference) |
| 12:00 pm | Lunch | |
| 12:45 | Perspectives from India | Sudeb Bhattacharya, Saha Institute of Nuclear Physics Kolkata (Calcutta), India |
| 1:45 | Accelerators and isotopes in industry/ medicine | Tom Ruth, Committee member |
| 2:45 | Break | |

*Closed Session*

## Thursday, September 23, 2010

*Closed Session*

## FOURTH MEETING
## IRVINE, CALIFORNIA
## FEBRUARY 12-13, 2011

## Saturday, February 12, 2011

*Closed Session*

## Sunday, February 13, 2011

*Closed Session*

# C

# Biographies of Committee Members

**Stuart J. Freedman**, *Chair*, was the Luis W. Alvarez Chair of Experimental Physics at the University of California at Berkeley with a joint appointment to the Nuclear Science Division of the Lawrence Berkeley National Laboratory. He received his Ph.D. from the University of California at Berkeley in 1972. His research experience spanned nuclear and atomic physics, neutrino physics, and small-scale experiments in particle physics, all focused on fundamental questions about the Standard Model. He co-chaired the American Physical Society (APS) physics of neutrinos study and the National Research Council's (NRC's) Rare Isotope Science Assessment Committee and served as a member of the NRC Committee on EPP2010: Elementary Particle Physics in the 21st Century. Dr. Freedman was a member of the National Academy of Sciences (NAS).

**Ani Aprahamian**, *Vice-Chair*, is a professor of experimental nuclear physics in the Department of Physics at the University of Notre Dame. She received her undergraduate and Ph.D. degrees from Clark University, Worcester, Massachusetts. Dr. Aprahamian's research focuses on the study of nuclear structure effects (shapes, masses, decay lifetimes, and probabilities) and how they can influence stellar processes. This research is a part of the new Joint Institute of Nuclear Astrophysics frontier center, established to address the fate of nuclei under extreme conditions such as accretion disks of binary neutron star systems or shock fronts of core collapse supernovae. The experiments are carried out by studying nuclei via radioactive ion beams at Notre Dame using the TWINSOL facility, the National Superconducting Cyclotron Laboratory (NSCL) facility at Michigan State University

(MSU), the Holifield Radioactive Ion Beam Facility (HRIBF) facility at Oak Ridge National Laboratory (ORNL), and the Argonne Tandem Linear Accelerator System (ATLAS) at Argonne National Laboratory (ANL). Dr. Aprahamian is co-chair of the Department of Energy (DOE) standing subcommittee on isotope production and applications of the Nuclear Science Advisory Committee (NSAC) and was the National Science Foundation (NSF) program director for nuclear physics and nuclear astrophysics. She is a fellow of the American Association for the Advancement of Science (AAAS) and the APS.

**Ricardo Alarcon** is a professor of physics at Arizona State University. He did his undergraduate studies at the University of Chile and received his Ph.D. in 1985 from Ohio University. He did postdoctoral work at the University of Illinois at Urbana-Champaign until 1989, when he joined Arizona State University as an assistant professor. His research covers experiments in electromagnetic nuclear physics and, more recently, in fundamental neutron science. He has held visiting professor appointments at the Massachusetts Institute of Technology (MIT) in 1995-1997 and 1999-2001 and served as project manager for the Bates Large Acceptance Spectrometer project at MIT-Bates from 1999 to 2002. He was a member of the DOE/NSF NSAC from 2001 to 2005. In 2003, he was elected a fellow of the APS. He was a member of the NRC Committee on Rare Isotope Science Assessment.

**Gordon A. Baym** is a professor of physics at the University of Illinois at Urbana-Champaign. He received his bachelor's degree in physics from Cornell University and his A.M. in mathematics and Ph.D. in physics from Harvard University. Dr. Baym has made seminal contributions to many fields, including developing much of the current understanding of the nature of neutron stars, relativistic effects in nuclear physics, condensed matter physics, quantum fluids, and most recently, Bose-Einstein condensates. He has written two textbooks on quantum mechanics and quantum statistical mechanics and has made major contributions to the scholarly study of the history of physics. Dr. Baym is a member of the NAS and the APS and was awarded the Hans A. Bethe Prize of the American Physical Society in 2002. He has participated in many activities for the NAS, NRC, and the Board on Physics and Astronomy (BPA), including serving as chair of the physics section of the NAS, participating in several decadal studies, and serving on the BPA governing board.

**Elizabeth Beise** is a professor of physics and interim associate provost for academic planning and programs at the University of Maryland in College Park. Her principal research interests in experimental nuclear physics focus on the use of electromagnetic and weak probes of the internal structure of protons, neutrons, and light nuclei, and on the use of nuclear physics techniques to test fundamental symmetries. She received the Maria Goeppert-Mayer Award from the APS in 1998 and is a fellow of both the APS (2002) and the AAAS (2009). From 2004 to 2006, she was a program director for nuclear physics at the NSF. She has served on several APS Division of Nuclear Physics committees, including its Executive and Program committees, as well as the APS Council and Executive Board. She was a member of the DOE-NSF NSAC (1999-2001) and was on the writing group for the NSAC Long Range Plans for Nuclear Science in 1996, 2002, and 2007.

**Richard F. Casten** is D. Allan Bromley Professor of Physics at Yale University. He received his Ph.D. from Yale in 1967. His field of research and expertise is nuclear structure physics and he has done both experimental and theoretical work. He brings knowledge of collective behavior and collective models in nuclei, the interacting boson approximation model, dynamical symmetries, quantum phase transitions and their critical point symmetries, the role of the proton-neutron interaction in the evolution of nuclear structure, exotic nuclei, and a large variety of experimental techniques. Dr. Casten was awarded the 2011 Tom W. Bonner Prize of the APS for his contributions to the study of regularities in nuclei and dynamical symmetries. He has honorary doctorates from the University of Bucharest and from Surrey University (U.K.), whose citation called him "one of the world's most distinguished nuclear physicists." He was the honoree at the "Mapping the Triangle" International Conference on Nuclear Structure in 2002 and is a fellow of the APS, the AAAS, and the Institute of Physics (IOP-U.K.) and an honorary fellow of the Hellenic Nuclear Physics Society. Dr. Casten was awarded the Senior (U.S.) Humboldt Prize and the 2009 Mentoring Award of the Division of Nuclear Physics (DNP) of the APS, primarily for mentoring women scientists throughout his career. He is in *Who's Who in the World* and has written a textbook, *Nuclear Structure from a Simple Perspective.*

He was director of the Wright Nuclear Structure Laboratory at Yale from 1995 to 2008. He was chair of NSAC in 2003, 2004, and 2005; chair of the DNP in 2008; a member of the DOE/NSF Long Range Planning Committee for Nuclear Science in 1989, 1995, 2001, and 2007; a member of the NRC Rare Isotope Sciences Assessment Committee (RISAC) in 2005 and 2006; chair of the Heavy Ion Research Center (GSI)-Facility for Antiproton and Ion Research (FAIR)-Nuclear Structure, Astrophysics, and Reactions (NuSTAR) Advisory Committee for the Future GSI Facility, 2004, 2005; chair, Science Advisory Committee for the Facility for Rare Isotope Beams (FRIB), 2009-2012; member the of International Union of Pure and Applied Physics (IUPAP) C-12 WG-9 subcommittee on international cooperation in nuclear physics, 2005-2008; cofounder and chair of the IsoSpin Laboratory Steering Group and Rare Isotope Accelerator Users Group, 1989-2004; member, FRIB Users Organization Executive Committee, 2008-2010; and member, Committee on International Perceptions of U.K. Research in Physics and Astronomy, 1999, 2005. He is the associate editor of *Physical Review C* for experimental nuclear structure, an editor for journals E and L of the *International Journal of Modern Physics* series, and is on the editorial board of *Nuclear Physics News.*

**Jolie A. Cizewski** is professor of physics in the Physics and Astronomy Department at Rutgers State University of New Jersey. She is an experimental physicist doing research in nuclear physics at the interface of nuclear structure, reactions, and astrophysics. She is also interested in the applications of nuclear physics to national nuclear security and in developing a talented and diverse workforce for national needs in nuclear science. Professor Cizewski is a fellow of the AAAS and the APS. She was also a recipient of a Sloan Foundation Fellowship and a Faculty Award for Women from the National Science Foundation. She served as the chair and member of the NRC Panel on Nuclear Data Compilations and as a member of the NRC Bits of Power Committee. She has served on NSAC, which advises DOE and the NSF, and was a member of the writing groups that developed the Long Range Plans for Nuclear Science in 2002 and 2007. She was also a coauthor of the 2004 NSAC report *Education in Nuclear Science.*

**Anna Hayes-Sterbenz** has been a staff member in the theoretical division at Los Alamos National Laboratory since 1997. Prior to joining Los Alamos, she was a staff member from 1991 to 1997 in theoretical nuclear physics at the Chalk River Nuclear Labs, in Canada. She has broad theoretical expertise in nuclear physics, spanning nuclear structure, neutrino-nucleus physics, fundamental symmetries, inertial confinement fusion, nonproliferation, and national defense. She is Principal Investigator (PI) and co-PI on a number of large-scale projects in both basic and applied nuclear physics and brings a perspective from the intersection of the two subfields. She is a fellow of the APS. She has served on several national committees, including the APS DNP executive committee, the organizing committee for APS DNP town hall meeting on the Nuclear Science Long Range Plan, the Los Alamos Neutron Science Center (LANSCE) advisory board, the Stockpile Science Academic Alliance DOE review committee, the Oak Ridge HRIBF Applications Working Group, and several NSF and DOE review and selection committees. At Los Alamos she chairs the Nuclear, Particle, Astrophysics and Cosmology (NPAC) Laboratory Directed Research and Development committee and is a member of the NPAC advisory team, the Los Alamos distinguished postdoc selection committee, and the Technical Working Group for the National Boost Initiative. She has also chaired the LANSCE User Group Executive Committee and served as the Theoretical Division nuclear weapons coordinator and the team leader for applied nuclear theory at Los Alamos.

**Roy J. Holt** is a distinguished fellow at Argonne National Laboratory, where he serves as chief of medium-energy research in the Physics Division. He is a distinguished experimentalist with broad expertise in low- and medium-energy nuclear physics. He brings knowledge of studies of light nuclei, the nucleon, and low-energy tests of fundamental symmetries. Dr. Holt is a fellow of the APS and of the IOP. He was the 2005 recipient of the Tom W. Bonner Prize of the APS for his experimental work in nuclear physics. He served on the NSAC subcommittee to generate the 2007 Long Range Plan in Nuclear Physics. During 1994-2000, he served as professor of physics at the University of Illinois at Urbana-Champaign, where he was also a director of the Nuclear Physics Laboratory. He has served on scientific program advisory committees for a number of accelerator facilities, including the Los Alamos Meson Physics Facility, the Stanford Linear Accelerator Center, MIT-Bates, the Indiana University Cyclotron Facility, and the Jefferson National Laboratory (JLAB) (chair); on NSAC subcommittees; and review panels for DOE, NSF, and the Natural Sciences and Engineering Research Council (NSERC) of Canada; and on the editorial boards of *Physical Review C, Nuclear Physics A,* and *Journal of Physics G.*

**Karlheinz Langanke** is the director of research at the GSI Helmholtz Zentrum für Schwerionenforschung in Darmstadt, Germany. He also is full professor at the Technische Universität Darmstadt and a senior fellow at the Frankfurt Institute of Advanced Studies. Before taking his current positions he held a chair for theoretical physics at Aarhus University in Denmark and has been a senior research associate at Caltech. His research expertise is in nuclear structure and reaction theory as well as in nuclear astrophysics. Dr. Langanke is supervisory editor for *Nuclear Physics A* and a member of the editorial boards of *Few Body Systems* and the *Atomic Data and Nuclear Data Tables.* Dr. Langanke serves on many advisory committees, including the International Science Advisory Committee of FRIB, to be constructed at the MSU site, and of the Institute of Physical and Chemical Research (RIKEN) in Tokyo. He has been the chairman of the Program Advisory Committee of TRIUMF (once known as the Tri-University Meson Facility) (Vancouver, Canada) and served on the program advisory committees of the Large Heavy Ion National Accelerator (GANIL) (France), the European Organization for Nuclear Research (CERN)/Isolde (Switzerland), RIKEN (Japan), and GSI (Darmstadt). He has been a member of the board of several international science institutions, including the INT in Seattle, NORDITA in Copenhagen, Denmark, the Joint Institute for Nuclear Astrophysics in the United States, and the European Center for Theoretical Nuclear Physics in Trento, Italy. Dr. Langanke has been a co-convener of the Nuclear Physics European Collaboration Committee (NuPECC) Long Range Plan in nuclear physics, written in 2003. During the last 16 months, Dr. Langanke has given lecture series on nuclear astrophysics at seven institutions and schools on four continents.

**Cherry A. Murray** (NAS/NAE) is dean of the Harvard School of Engineering and Applied Sciences and serves as chair of the NRC's Division on Engineering and

Physical Sciences. Her research interests are the physics of surfaces, condensed matter, and complex fluids, with an emphasis on light scattering and imaging. In addition to her research, Dr. Murray has substantial background in research management, having served as deputy director for science and technology at the LLNL, after serving as senior vice president for Bell Labs Research, Lucent Technologies.

**Witold Nazarewicz** is a professor of physics at the Department of Physics, University of Tennessee, and distinguished R&D staff at the Physics Division, Oak Ridge National Laboratory. He is a distinguished theorist with broad expertise in nuclear physics, many-body problems, interdisciplinary many-body science, and computational physics. He is listed by the Institute for Scientific Information as among the most highly cited authors in physics. Dr. Nazarewicz is a fellow of AAAS, APS, and IOP. He was awarded the 2012 Tom W. Bonner Prize of the APS for his work in developing and applying nuclear density functional theory, motivating experiments and interpreting their results, and implementing a comprehensive theoretical framework for the physics of exotic nuclei. He received an honorary doctorate from the University of the West of Scotland and previously served on two NRC committees—the Committee on Nuclear Physics (1996-1999) and the Rare Isotope Science Assessment Committee (2005-2007). He is a director of the Universal Nuclear Energy Density Functional (UNEDF) Scientific Discovery through Advanced Computing (SciDAC) Program at DOE, associate editor of *Reviews of Modern Physics*, editor with *Computer Physics Communications*, member of the FRIB Science Advisory Committee, and member of the steering committees of the Japan-U.S. Theory Institute for Physics with Exotic Nuclei (JUSTIPEN) and the France-U.S. Theory Institute for Physics with Exotic Nuclei (FUSTIPEN). Dr. Nazarewicz has served on numerous DOE, NSF, and DNP/APS committees, including NSAC; was a member of the nuclear physics Long-Range Planning Working Groups in 1995, 2001-2002, 2005, and 2007-2008; and has served on advisory committees of the National Superconducting Cyclotron Laboratory (NSCL)/MSU, ATLAS/ANL, the 88-in. cyclotron at Lawrence Berkeley National Laboratory (LBNL), HRIBF/ORNL, Institute for Nuclear Theory/Seattle, JLAB, and TRIUMF/Canada. In 2000-2005 he was a co-chair and chair of the RIA Users Organization.

**Konstantinos Orginos** is an assistant professor of physics at the College of William and Mary. He is also a senior staff member of the theory center at the JLAB. He received his Ph.D. in physics from Brown University in 1998, worked as a postdoctoral associate at the University of Arizona and the Brookhaven National Laboratory, and was a research scientist at the Laboratory for Nuclear Science at MIT. He joined the faculty at the College of William and Mary in 2005. He is the recipient of a DOE Outstanding Junior Investigator award. His research focuses on lattice quantum chromodynamics (QCD) calculations relevant for understanding the structure of hadrons and the emergence of the nuclear force from QCD. He has broad experience on the use of high-performance computing for performing calculations relevant to hadronic structure and interactions relevant for establishing the connection between QCD and nuclear physics.

**Krishna Rajagopal** is a professor of physics at MIT and is the associate head for education of the MIT Department of Physics. He obtained his doctorate at Princeton in 1993 and then spent 3 years at Harvard as a junior fellow and 1 year at Caltech before coming to MIT in 1997. Dr. Rajagopal enjoys thinking about QCD in extreme conditions because it requires linking usually disparate strands of theoretical physics, including nuclear physics, particle physics, string theory, condensed matter physics, and astrophysics. His research interests include the properties of the cold dense quark matter that may lie at the centers of neutron stars. His work shows that this stuff is a transparent insulator, not an electric conductor as previously assumed, and may in a certain sense be crystalline. Dr. Rajagopal also studies the hot quark soup that filled the universe shortly after the big bang and that is created in current experiments at the Relativistic Heavy Ion Collider (RHIC). He uses gauge/gravity duality—originally developed by string theorists trying to understand quantum gravity—to understand properties of hot quark soup. He has also analyzed the critical point in the QCD phase diagram and has proposed signatures for its experimental detection. Dr. Rajagopal serves on the RHIC Program Advisory Committee and the editorial board of *Physical Review D*. He is a member of the executive committee of the DNP of the APS. He served on the NSAC subcommittee on nuclear theory. He is a fellow of the APS. He was the MIT Class of 1958 assistant professor and has been a DOE Outstanding Junior Investigator and an Alfred P. Sloan research fellow.

**R.G. Hamish Robertson** is the Boeing Distinguished Professor of Physics at the University of Washington and director of the Center for Experimental Nuclear Physics and Astrophysics. He took his undergraduate degree at Oxford and his Ph.D. in atomic-beam and nuclear-structure physics at McMaster. Upon graduation, Dr. Robertson went to Michigan State University as a postdoctoral fellow and remained on the faculty, becoming a professor of physics in 1981. In that same year, he joined Los Alamos National Laboratory (LANL) and investigated neutrino mass via tritium beta decay and solar neutrino physics. Dr. Robertson was appointed a fellow of LANL in 1988 and initiated the laboratory's collaboration in the Sudbury Neutrino Observatory project. He has served as the U.S. co-spokesman for that project and was scientific director in 2003-2004. Results from this experiment have shown that neutrinos have mass and are strongly mixed in flavor, in contradiction to the Standard Model of particle physics. In 1994, Dr. Robertson took a professorship at the University of Washington, where he is continuing his work in neutrino physics. In 2003 he was elected to fellowship in the American Academy of Arts and Sciences and in 2004 to the NAS. A past member of NRC's Board of Physics and Astronomy, he has also served on several committees, including the Nuclear Physics and Neutrino Astrophysics panels.

**Thomas J. Ruth** is senior research scientist at TRIUMF and senior scientist at the British Columbia Cancer Research Centre. In addition he is adjunct professor of pharmaceutical sciences and medicine at the University of British Columbia, of chemistry at Simon Fraser University, and of physics at the University of Victoria. He is a leader in the production and application of radioisotopes for research in the physical and biological sciences. His efforts at establishing positron emission tomography (PET) as a quantitative tool for in vivo biochemistry have been recognized by the Canadian Nuclear Medicine Society's highest award of meritorious status. He has served on a multitude of committees, including the Institute of Medicine's (IOM's) Committee on Medical Isotopes (1995), the NRC's Committee on the State of the Science in Nuclear Medicine, the IOM panel on the Status and Future of Nuclear Medicine (2007-2008), and the NAS panel on the Production of Medical Isotopes without HEU (2008-2009). He was a member of the NSAC Subcommittee on Isotopes (2009). In addition he serves as an expert on radioisotope production for the International Atomic Energy Agency. He has published more than 250 peer-reviewed papers and book chapters. Dr. Ruth received his Ph.D. in nuclear spectroscopy from Clark University.

**Hendrik Schatz** is professor of physics at the Department of Physics and Astronomy and the National Superconducting Cyclotron Laboratory at Michigan State University. He is associate director and cofounder of the Joint Institute for Nuclear Astrophysics, an NSF Physics Frontiers Center. Dr. Schatz is a distinguished experimentalist who works at the intersection of nuclear physics and astrophysics and has also contributed to the theoretical understanding of nuclear processes in the cosmos. His particular interests are rare isotope beam experiments and the application of the results to explosive stellar processes and neutron stars. Dr. Schatz is a fellow of the APS and a member of the NSAC. He has co-chaired with Robert Janssens the town meeting "Study of Nuclei and Nuclear Astrophysics" (including co-authorship of the associated white paper) and was a member of the writing committee for the NSAC 2007 Nuclear Physics Long Range Plan. He is also a member of the NRC Stars and Stellar Evolution Panel, one of the five science frontier panels of the decadal survey of astronomy and astrophysics (Astro2010). He has given an invited presentation to the NRC Rare Isotope Science Assessment Committee (RISAC) on rare isotope studies for nuclear astrophysics. Dr. Schatz is a member of the Science Advisory Committee for the FRIB and member of the program advisory committee at the ATLAS facility at Argonne National Laboratory and the GSI rare isotope facility in Germany.

**Robert E. Tribble** is Distinguished Professor of Physics and Astronomy and director of the Cyclotron Institute at Texas A&M University. He is an international leader in experimental nuclear physics and nuclear astrophysics. His seminal contributions both in instrument development and in measurement techniques have led him—and the many researchers around the world who have copied his methods—to important new understanding of the fusion reactions that occur in stars and stellar explosions. In addition, he has made key contributions to the search for physics beyond the Standard Model of particle physics and has also played a leading role in a large-scale experiment that studied the quark composition of the proton. He recently completed a 3-year term as chair of the NSAC. The most recent NSAC Long Range Plan for nuclear science was completed during his tenure. He has served on numerous NSF and DOE review panels and NSAC subcommittees, including the 2005 NSAC subcommittee on implementing the 2002 Long Range Plan, which he chaired. He is presently either a member or chair of four program advisory committees for facilities around the world and a member of science advisory committees for ANL, JLAB, the FRIB, and a new national laboratory in Korea based on the KORIA accelerator facility. Last year he was voted vice-chair of the APS Division of Nuclear Physics.

**William A. Zajc** is a professor and chair of the Physics Department at Columbia University. His undergraduate studies were at the California Institute of Technology, and he obtained his Ph.D. in physics at the University of California at Berkeley. Dr. Zajc's research interests center on the experimental study of QCD as studied via collisions of nuclei at relativistic energies, with an emphasis on understanding the properties of matter under the extreme conditions where the quark and gluons no longer are confined to individual neutrons and protons. From 1997 to 2006 Dr. Zajc served as spokesperson for the PHENIX experiment (Pioneering High Energy Nuclear Interaction Experiment) at BNL's RHIC and was deeply involved in the 2005 discovery of "the perfect liquid" formed in collisions of heavy nuclei at RHIC. Among his many professional activities are participation in the 1996, 2002, and 2007 NSAC Long Range Plan writing groups and service on the NSAC (2004-2007). He has also served on the advisory committee of the Institute for Nuclear Theory, the editorial board of *Annual Reviews of Nuclear and Particle Physics* and the JLAB Science Council. He is currently vice-chair of the BNL Science and Technology Steering Committee. In 2010, Dr. Zajc served as chair of the APS DNP (2010-2011). Dr. Zajc is a fellow of the APS and of the AAAS.

# D

# Acronyms

| | |
|---|---|
| AAAS | American Association for the Advancement of Science |
| AD | antiproton decelerator |
| AEGIS | Antihydrogen Experiment Gravity Interferometry Spectroscopy |
| AGATA | Advanced Gamma Tracking Array |
| AGS | Alternating Gradient Synchrotron |
| ALICE | A Large Ion Collider Experiment |
| ALTO | Accélérateur Linéaire auprès du Tandem d'Orsay (Linear Accelerator Near the Tandem of Orsay) |
| AMS | accelerator mass spectrometry |
| ANDES | Agua Negra Deep Experiment Site |
| ANDESLab | Agua Negra Deep Experiment Site Lab |
| ANL | Argonne National Laboratory |
| APCTP | Asian Pacific Center for Theoretical Physics |
| APPA | Atomic, Plasma Physics and Applications (collaboration) |
| APS | American Physical Society |
| ARIEL | Advanced Rare Isotope Laboratory |
| ATLAS | Argonne Tandem Linear Accelerator System |
| ATTA | atom trap trace analysis |
| AURA | Association of Universities for Research in Astronomy |
| AVF | azimuthal varying field |
| | |
| BaBar | B and B-bar experiment |

| | |
|---|---|
| BEC | Bose-Einstein condensate |
| BEPCII | Beijing Electron Positron Collider II |
| BES-III | Beijing Spectrometer III |
| BGO | bismuth orthogermanate |
| BNL | Brookhaven National Laboratory |
| BPA | Board on Physics and Astronomy |
| BRIF | Beijing Rare Ion Beam Facility |
| BSI | Basic Science Institute |
| | |
| CAREER | Faculty Early Career Development (CAREER) Program (NSF) |
| CBM | Compressed Baryonic Matter |
| CEBAF | Continuous Electron Beam Accelerator Facility |
| CERN | Conseil Européen pour la Recherche Nucléaire (European Organization for Nuclear Research) |
| CEU | Conference Experience for Undergraduates |
| CGC | color glass condensate |
| CI | configuration interaction |
| CKM | Cabibbo-Kobayashi-Maskawa |
| CLEO | Short for "Cleopatra" (a general-purpose particle detector at the Cornell Electron Storage Ring |
| CMS | Compact Muon Solenoid |
| CNEA | Comisión Nacional de Energía Atómica (National Atomic Energy Commission) |
| COMPASS | Common Muon and Proton Apparatus for Structure and Spectroscopy |
| COSY | Cooler Sychrotron facility (Germany) |
| CP | combination of C-symmetry and P-symmetry |
| CT | computed tomography |
| CUORE | Cryogenic Underground Observatory for Rare Events |
| CXC | Chandra X-Ray Observatory |
| CZT | cadmium zinc telluride |
| | |
| DANCE | Detector for Advanced Neutron Capture Experiments |
| DESIR | Désintégration, Excitation et Stockage d'Ions Radioactifs (disintegration, excitation, and storage of radioactive ions) |
| DESY | Deutsches Elektronen Synchrotron (German Electron Synchrotron) |
| DFT | density functional theory |
| DNP | Division of Nuclear Physics |
| DOE | Department of Energy |

DUSEL            Deep Underground Science and Engineering Laboratory

EBIS             Electron Beam Ion Source
ECT*             European Center for Nuclear Theory and Related Areas
EDM              electric dipole moment
EFT              effective field theory
EIC              electron-ion collider
ELENA            Extra Low Energy Antiproton Ring
ELI-NP           Extreme Light Infrastructure
ELSA             Electron Stretcher and Accelerator
EMC              European Muon Collaboration
EMMI             Extreme Matter Institute
EOS              equation of state
ESA              European Space Agency
ESR              experimental storage ring
EURISOL          European ISOL facility
EXO              Enriched Xenon Observatory

FACA             Federal Advisory Committee Act
FAIR             Facility for Antiproton and Ion Research
FDG              fluorodeoxyglucose
FEL              free-electron laser
FLAIR            Facility for Low-Energy Antiproton and Ion Research
FNAL             Fermi National Accelerator Laboratory
FNPB             Fundamental Neutron Physics Beamline
FRIB             Facility for Rare Isotope Beams
FUSTIPEN         French-U.S. Theory Institute for Physics with Exotic Nuclei

GANIL            Grand Accélérateur National d'Ions Lourds (Large Heavy Ion
                   National Accelerator)
GlueX            gluon experiment at JLAB
GRETA            Gamma Ray Energy Tracking Array
GRETINA          Gamma Ray Energy Tracking In-Beam Nuclear Array
GSFC             Goddard Space Flight Center
GSI              Gesellschaft für Schwerionenforschung (Heavy Ion Research
                   Center, now the GSI Helmholtz Centre for Heavy Ion
                   Research GmbH)

HERA             Hadron-Elektron-Ring-Anlage
HEU              highly enriched uranium
HIE-ISOLDE       High Intensity and Energy ISOLDE

| | |
|---|---|
| HIGS | High Intensity Gamma Source |
| HIRFL-CSR | Heavy Ion Research Facility in Lanzhou-Cooler Storage Ring |
| HPGe | high-purity germanium |
| HRIBF | Holifield Radioactive Ion Beam Facility |
| HST | Hubble Space Telescope |
| | |
| IAEA | International Atomic Energy Agency |
| ICF | inertial confinement fusion |
| IJMPE | International Journal of Modern Physics-E |
| ILL | Institut Laue-Langevin |
| INO | India-based Neutron Observatory |
| INT | Institute for Nuclear Theory |
| IR | infrared |
| ISAC-I | Isotope Separator and Accelerator (part 1) |
| ISAC-II | Isotope Separator and Accelerator (part 2) |
| ISOL | Isotope Separator Online |
| ISOLDE | Isotope Separator Online Detector |
| ITER | International Thermonuclear Experimental Reactor |
| IUPAP | International Union for Pure and Applied Physics |
| | |
| JILA | Joint Institute for Laboratory Astrophysics |
| JINA | Joint Institute for Nuclear Astrophysics |
| JINR | Joint Institute for Nuclear Research |
| JLAB | Thomas Jefferson National Accelerator Facility |
| J-PARC | Japan Proton Accelerator Research Complex |
| JPL | Jet Propulsion Laboratory |
| JPS | Japanese Physical Society |
| JSC | Jülich Supercomputing Center |
| JUGENE | Jülich Blue Gene |
| JUSEIPEN | Japan-U.S. Experimental Institute for Physics with Exotic Nuclei |
| JUSTIPEN | Japan-U.S. Theory Institute for Physics with Exotic Nuclei |
| | |
| KamLAND | KAMioka Liquid Scintillator Antineutrino Detector |
| KamLAND-Xen | KAMioka Liquid Scintillator Antineutrino Detector-Xenon |
| KATRIN | KArlsruhe TRItium Neutrino |
| KEI or K | Japanese Supercomputer (named for word "kei," meaning 10 quadrillion) |
| KEK | Kō Enerugī Kasokuki Kenkyū Kikō (High Energy Accelerator Research Organization) |
| KoRia | Korea Rare Isotope Accelerator |

| | |
|---|---|
| LANL | Los Alamos National Laboratory |
| LANSCE | Los Alamos Neutron Science Center |
| LBNL | Lawrence Berkeley National Laboratory |
| LHC | Large Hadron Collider |
| LIGO | Laser Interferometer Gravitational-Wave Observatory |
| LINAC | linear particle accelerator |
| LLC | low-level counting |
| LLNL | Lawrence Livermore National Laboratory |
| LMA | large mixing angle |
| LRP | long-range plan |
| LSDS | Lead Slowing-Down Spectrometer |
| LSO | lutetium orthosilicate |
| LUNA | Laboratory for Underground Nuclear Physics |
| | |
| MAMI | Mainz Microtron |
| MAX-lab | Microtron Accelerator for X-ray Production (National Electron Accelerator Laboratory for Synchrotron Radiation Research, Nuclear Physics, and Accelerator Physics) |
| MEGa-Ray | Mono-Energetic Gamma-Ray facility (LLNL) |
| MERCOSUR | Mercado Común del Sur (Common Southern Market) |
| MIT | Massachusetts Institute of Technology |
| MoNA | Modular Neutron Array |
| MOX | mixed oxide fuel |
| MRI | magnetic resonance image/imaging |
| MSU | Michigan State University |
| MTAS | Modular Total Absorption Spectrometer |
| | |
| N | neutron number |
| NASA | National Aeronautics and Space Administration |
| NESR | new ESR |
| NFS | Neutron for Science |
| NIF | National Ignition Facility |
| NIH | National Insitutes of Health |
| NIST | National Institute of Standards and Technology |
| NNSA | National Nuclear Security Administration |
| NORDITA | Nordic Institute for Theoretical Physics, aka Nordic Institute for Theoretical and Atomic Physics |
| NPAC | nuclear, particle, astrophysics and cosmology |
| NPT | Nuclear Non-Proliferation Treaty |
| NRC | National Research Council |
| NSAC | Nuclear Science Advisory Committee |

| NSCL | National Superconducting Cyclotron Laboratory |
| NSERC | Natural Sciences and Engineering Research Council of Canada |
| NSF | National Science Foundation |
| NSM | New Standard Model |
| NuPECC | Nuclear Physics European Collaboration Committee |
| NuSTAR | Nuclear Structure, Astrophysics and Reactions |
| | |
| OECD | Organisation for Economic Co-operation and Development |
| ORNL | Oak Ridge National Laboratory |
| ORRUBA | Oak Ridge–Rutgers University Barrel Array |
| | |
| PANDA | Antiproton Annihilation at Darmstadt |
| PET | positron emission tomography |
| PHENIX | Pioneering High-Energy Nuclear Interaction Experiment |
| PI | principal investigator |
| PMT | photomultiplier tube |
| POPA | Panel on Public Affairs |
| PPC | p-type point contact |
| PREX | Lead Radius Experiment |
| PSI | Paul Scherrer Institut |
| PV | parity violating |
| | |
| QCD | quantum chromodynamics |
| QCDOC | QCD on a chip |
| QED | quantum electrodynamics |
| QGP | quark-gluon plasma |
| $Q_s$ | gluon momentum scale below which saturation is thought to arise |
| | |
| RBRC | RIKEN-BNL Research Center |
| RCNP | Research Center for Nuclear Physics |
| REX-ISOLDE | Radioactive Beam Experiment at ISOLDE |
| RHIC | Relativistic Heavy Ion Collider |
| RIB | radioactive ion beam |
| RIBF | Radioactive Ion Beam Facility |
| RIBRAS | Radioactive Ion Beams in Brasil |
| RIKEN | Rikagaku Kenkyujo (Institute of Physical and Chemical Research) |
| RISAC | Rare Isotope Science Assessment Committee |